P9-DFP-624

ANALYTICAL PROBLEMS IN BIOLOGY

Michael P. Donovan
Robert D. Allen

West Virginia University
Morgantown, West Virginia

David CHavez-Gonzalez
P.C.C fall 2010

PEARSON
Custom
Publishing

Copyright © 1983 by **Burgess Publishing Company**

Printed in the United States of America

ISBN 0-8087-9241-5

All rights reserved. No part of this book may be reproduced in any form whatsoever, by photocopy, xerography, or photograph or by any other means, by broadcast or transmission, by translation into any kind of language, nor by recording electronically or otherwise, without permission from the publisher, except by a reviewer, who may quote brief passages in critical articles and reviews.

20 19 18 17

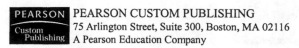
PEARSON CUSTOM PUBLISHING
75 Arlington Street, Suite 300, Boston, MA 02116
A Pearson Education Company

ACKNOWLEDGEMENTS

All of the instructors in the General Biology Program at West Virginia University have contributed ideas and words to this book. We thank them for their efforts and cooperation. They are:

Arnold Benson	Anthony Janicki
Roy B. Clarkson	Joseph Marshall
Jesse F. Clovis	Michael B. Moll
M. O. Coover	Mark J. Pettenatti
Dorothy C. Dunning	William Pietraface
Sharon Erickson	Dennis C. Quinlan
Ramsey H. Frist	Martin W. Schein
Willis H. Hertig, Jr.	Walter R. Statkiewicz
William H. Hunt	David Stroup
Henry W. Hurlbutt	Gerald R. Wilcox

Leah A. Williams

We give special thanks to Carol Hando and Gloria Allen who processed all these words and produced the final typescript.

CONTENTS

INTRODUCTION

The problems in this book require students to apply the fundamental concepts of biology in an analytical way. These are not memory problems and, in general, the "correct" answers cannot be found explicitly in a textbook. The students must use their knowledge of one or several concepts to solve a problem which represents a new or novel application. As a consequence the best answer to a problem may not be immediately obvious and reaching the best solution requires careful thought and analysis.

Most of the problems address points which are particularly difficult for our students. Although the concepts used are fundamental to any introductory biology course, and students can be expected to know these concepts, their application even to apparently simple problems can be quite challenging.

Each problem is presented in a multiple-choice format. For maximum benefit each choice should be considered individually and an argument should be written for accepting or rejecting it. Since each problem has one best answer there should be one argument for acceptance and four for rejection. Examples of problems solved in this way are shown in each chapter.

Several of the choices in a problem may be true statements but may not apply to that particular problem. Picking the "right" answer is not sufficient; students must be able to justify their choice of the best answer with a rational argument. If used properly, these problems can be a decided aid in developing analytical thinking.

Throughout the book strong emphasis is placed on the process of science: identifying assumptions, drawing conclusions, making predictions, and evaluating evidence. We have found that skill with these operations develops slowly and requires considerable practice. We have accordingly included problems which require practice with these operations in all chapters.

BIOLOGY: A SCIENCE AND ITS METHODS

The science of biology is a way of generating explanations of the living world. It begins with observations, precise and accurate information about the properties and actions of organisms. Tentative explanations of observations are called hypotheses. Each hypothesis leads to predictions about the results to be expected when something is done to living things. If two or more hypotheses make different predictions about the same action, carrying out the action is an experiment. Comparing the observed results to the predictions allows a choice to be made among the hypotheses; a conclusion can be drawn. A good experimental design includes proper controls, elimination of bias, sufficient sample size and adequate repetition. Making hypotheses, predictions and conclusions is collectively termed interpretation of observations. Interpretation invariably requires accepting some statements as true, even though there is no evidence to support them. Such statements are called assumptions.

Organisms are made of molecules which are formed and interconverted by chemical reactions. Reactions of biochemical molecules rarely proceed at appreciable rates unless appropriate catalysts are present. In organisms the catalysts are proteins called enzymes. Most of the mass of organisms, aside from water, consists of polymers, chains of similar small units called monomers.

The basic functional unit into which biological molecules are organized is the cell. Cells may be observed with the light microscope when they are living or after extensive processing to improve their visibility. Living material cannot be observed in the transmission electron microscope and only a few living organisms can be observed in the scanning electron microscope, but these two instruments provide higher magnification and resolution and different information than the light microscope. Cells are also studied by cell fractionation, a process of disrupting cells by homogenization and separating their components by centrifugation. Cell fractionation and the preparation of cells of microscopic study are both complicated and drastic physical and chemical processes. Thus there is a danger that they will cause artifacts, properties that do not exist in the living cell.

EXAMPLE 1.

Suppose a marine biologist is diving near the seashore when he observes a gray fish resting on gray sand on the ocean floor. Later he notices another fish resting on the ocean floor in an area composed of small, variously colored stones. This second fish is identical to the first except that it is spotted instead of gray. The biologist continues his search of the area and collects 15 of the gray fish and 11 of the spotted fish. After taking the fish back to his laboratory he consults a reference book and discovers that the fish are identified as two separate species. He then places all the fish he collected in a large aquarium with gray sand on the bottom. Several hours later he observes that all the fish are gray. Which of the following would be the LEAST acceptable statement about these findings?

1) Gray fish tend to rest on gray background.
2) The reference book was mistaken in identifying the two kinds of fish as separate species based on color.
3) The biologist had happened upon a small population of gray fish which could become spotted.
4) Some fish can change color when they move to backgrounds of a different color.
5) The biologist had happened upon a small population of spotted fish which become gray.

ANALYSIS:

1. Since all the gray fish collected by the biologist were found resting on gray sand this statement would be highly acceptable.

2. The biologist has good evidence that the fish are the same; his observations indicate that spotted fish can turn gray and are thus identical in all respects to the gray fish collected from the ocean. Unless the scientist or scientists writing the reference book had observed a color change by these fish, they may well have considered them different kinds of fish. Acceptable statement.

3. Although this statement would seem acceptable (if spotted fish can turn gray then gray fish would likely be able to turn spotted), the biologist has no direct observations indicating gray fish can turn spotted. Therefore, least acceptable.

4. His observations suggest some fish can change color when placed on a different background (spotted fish can turn gray). Acceptable statement.

5. His observations are only on a small population of spotted fish. The observed behavior may not be characteristic of all spotted fish, although this possibility would seem remote. Acceptable statement.

EXAMPLE 2.

A biochemist was studying the combination of two reactants to form a single product. An enzyme catalyzed the reaction. As he increased the temperature, the speed of the reaction increased until a certain temperature was reached. If the temperature was raised further, the speed of the reaction decreased rapidly. Which of the following is the best conclusion to be drawn from these results?

1) As temperature is increased, the reactants move more rapidly.
2) As temperature is increased, the products move more rapidly.
3) Above a certain temperature the enzyme combines better with the reactants.
4) Above a certain temperature the enzyme begins to move.
5) Above a certain temperature the shape of the enzyme is changed drastically.

4

ANALYSIS:

1. If the reactants move more rapidly, the frequency of collisions should increase and the rate of reaction should increase. This is observed but only up to a point. This mechanism does not account for the rapid decrease in reaction rate at high temperatures. This mechanism predicts a continued increase in reaction rate at high temperature.

2. If the products move more rapidly, the rate of reaction should not change. Collisions between reactants influence reaction rate. This is unaffected by the velocity of product molecules. Like choice one this choice is probably true, but it does not account for all the observations.

3. An enzyme-catalyzed reaction occurs when reactants combine appropriately with enzyme. If the enzyme combines better with reactants under some conditions, the reaction should therefore go faster. We observe that it goes slower.

4. If the enzyme begins to move, it should undergo more collisions with reactants and the reaction should go faster.

5. When reactants combine with an enzyme, they are held in proper orientation on the enzyme's surface so that a reaction may occur between them. If the shape of the enzyme changes, the fit between reactant and enzyme will not be as good, resulting in less effective orientation and reduced reaction rate. This accounts for the observed decrease in reaction rate.

1. One day you meet a student watching a wasp drag a paralyzed grasshopper down a small hole in the ground. When asked what he is doing he replies, "I'm watching that wasp store paralyzed grasshoppers in her nest to feed her offspring." Which of the following is the best description of his reply?

    1) He is not a careful observer.
    2) He is stating a conclusion only partly derived from his observation.
    3) He is stating a conclusion entirely drawn from his observation.
    4) He is stating his observation.
    5) He is making no assumptions.

2. Which of the following additional observations would add the most strength to the student's reply in Question 1?

    1) Observing the wasp digging a similar hole.
    2) Observing the wasp dragging more grasshoppers into the hole.
    3) Observing the wasp sealing the entrance to the hole and concealing its location.
    4) Digging into the hole and observing wasp eggs on the paralyzed grasshoppers.
    5) Observing adult wasps emerging from the hole a month later.

3. Both of you wait until the wasp leaves the area, then you dig into the hole and observe three paralyzed grasshoppers, each with a white egg on its side. The student states that this evidence supports his reply in Question 1. Which of the following assumptions is he making?

    1) The eggs are grasshopper eggs.
    2) The wasp laid the eggs.
    3) The wasp dug the hole.
    4) The wasp will return with another grasshopper.
    5) The wasp has left permanently.

4. You take the white eggs to the Biology laboratory. Ten days later immature wasps hatch from the eggs. The student states that this evidence supports his reply in Question 1. Which of the following assumptions is he making?

    1) The wasp dug the hole.
    2) The wasp stung the grasshoppers.
    3) The grasshoppers were dead.
    4) The grasshoppers will not decay in ten days or less.
    5) A paralyzed grasshopper cannot lay an egg.

5. A biology textbook states that one of the shortcomings of a diagram of a "typical cell" is that it "mixes the information obtained by different procedures, such as electron microscopy and light microscopy." Why is this a shortcoming?

    1) You would never see a cell that looked like the diagram by using any one available technique.
    2) Using different techniques would mean that the diagram was a composite of information from more than one cell.
    3) Some techniques are known to produce artifacts.
    4) Different techniques are likely to produce different artifacts and lead to incompatible misinterpretations.
    5) It is not a shortcoming because it results in a more complete representation of any particular cell.

6. The first sentence of a recent scientific paper says, "Reproduction in many species is confined to a time of year when the probability of survival for both the adults and offspring is

maximum". When you start reading this paper you would be most justified in saying that this statement is:

1) A fact.
2) An observation.
3) An interpretation based on observations.
4) An assumption based on controlled experiment.
5) False.

7. Two experimenters performed the same experiment. After a discussion they discovered that their conclusions did not agree. If you were asked to resolve their problem how would you best proceed?

1) Examine their experimental designs and their data.
2) Repeat the experiment yourself.
3) Have the experimenters repeat their experiments.
4) Examine the literature that reports the results of their experiment.
5) Ask an authority in that area what the answer is.

8. Suppose as a result of your investigation in Question 7 you suspect that one of the experimenters was biased during the course of the experiment. If you accept the other experimenter's conclusion, which of the following assumptions did you make?

1) No conclusion should be accepted without evidence.
2) If the experimenter is biased, his conclusion is not supported.
3) A conclusion produced by a biased experimenter may still be valid.
4) The unbiased experiment may have resulted in an unsupported conclusion.
5) The biased conclusion should be accepted.

9. A student studying cricket calls hypothesized that a call of normal loudness would be more attractive to females than a call twice as loud. He devised a test in which he released a female in the middle of a long tube. At the left end of the tube was a speaker playing back the sounds at normal loudness and at the right end of the tube was a speaker playing back the calls at twice the normal loudness. In all 20 trials the female cricket approached the speaker playing back the sounds at normal loudness. Which of the following is the best statement regarding this experiment?

1) The female was "left-handed" and so always went left.
2) The normal sound attracted the female.
3) The sound that was twice as loud repelled the female and made her move toward the sound of normal loudness.
4) The design of the experiment did not adequately test the hypothesis.
5) The design of the experiment was satisfactory but the results are suspect because only one female was used.

10. To test the hypothesis that *Callinectes* crabs actively avoid light, a student placed 20 crabs in an aquarium in which 24 shelters were located. Twelve of the shelters were made of glass and 12 others were made of rocks. At the end of the hour the student noted the distribution of the crabs. He found 8 under glass and 12 under rocks. He concluded that the crabs showed NO preference for light or dark shelters and the hypothesis should be rejected. What comment best describes his design?

1) He did not give the crabs enough time to adjust to finding new shelters.
2) His design did not test his hypothesis.
3) He did not control for time of day or tide.
4) He did not specify the sex of his crabs.
5) The design adequately tested his hypothesis.

11.　In which of the following cases are you most likely acting because you suspect your roommate is biased?

　　1)　You don't buy a certain brand of stereo after your roommate, who owns one, plays it for you.
　　2)　You do buy a certain brand of stereo even though your roommate says it is poor.
　　3)　You do buy a certain brand of stereo which your roommate recommends. His father sells it to you at a discount.
　　4)　1 and 2 are equally likely.
　　5)　2 and 3 are equally likely.

12.　A scientist reported a study of the effects of a certain drug on 1000 volunteer subjects. He concluded that the drug had no toxic effects on humans. A student read the report and suspected that the scientist was biased. Which of the following observations would be most likely to make a student suspect that the scientist was biased?

　　1)　Another paper reached the opposite conclusion.
　　2)　The Results section of the paper indicated that 952 subjects were examined at the end of the study but 48 were not.
　　3)　The manufacturers of the drug referred to the study in their application to the Food and Drug Administration.
　　4)　The Discussion section of the paper admitted that the scientist assumed that all of the subjects were healthy and had never taken this drug before.
　　5)　The scientists only gave the drug to half of the subjects.

13.　Independent movement of an organism towards increasing concentrations of a chemical substance is called positive chemotaxis. A student devises a way to measure chemotaxis of bacteria. He places a porous filter between a solution of the chemical to be tested and a standard suspension of bacteria. The experiment is duplicated, but without any test chemical in the solution. After a fixed time the filters are removed and the number of bacteria trapped in each filter is determined. If significantly more bacteria are found in the filter when the test chemical is present than when it is not present, the student scores the chemical solution as an active agent that induces chemotaxis. Which of the following statements is the most valid criticism of the student's experiment?

　　1)　The experiment cannot be repeated.
　　2)　The filter might have attracted the bacteria.
　　3)　The chemical could have stimulated random movement of the bacteria cells.
　　4)　Gravity may have caused the movement.
　　5)　The test solution may have contained impurities.

14.　A biologist wanted to test the hypothesis that ants lay a trail of chemicals as they walk and are able to find their way back to the starting point by following the trail. He observed some ants crossing a table to and from a food source and he rubbed his finger across their path. They began to walk aimlessly in loops and could not find the food. He concluded that his hypothesis was supported. Which of the following is the best criticism of his experiment and his conclusion?

　　1)　The hypothesis is untestable.
　　2)　He only did the experiment with one kind of ant.
　　3)　He did not test the effect of light coming from only one direction.
　　4)　There is an alternate interpretation of his observation.
　　5)　He should have rubbed a dead ant across the path.

15.　A prominent biology textbook described a study of two species of duckweed, a very small plant that floats in or on water. "Grown alone in a tank, each did well but when they were grown in the same tank, *Lymna polyrhiza* died out. The reason is, *L. gibba* is a better competitor for light. It has air-filled sacs that cause it to float higher in the water, blocking light from its competitor." Which of the following observations would add the most strength to their conclusion?

1) *L. polyrhiza* does not survive in water from a tank where *L. gibba* is growing.
2) *L. polyrhiza* does survive in water from a tank where *L. gibba* is growing, if it is filtered first.
3) Both species survive in a tank illuminated from the bottom only.
4) *L. gibba* dies out when both species are placed in a tank illuminated from the bottom only.
5) Both 1 and 2.

16. The following data on weight gain of a certain kind of fish were obtained when the fish were grown at various temperatures for one month.

Temperature	Weight Gain
5°C	70 gm
15°C	142 gm
25°C	293 gm

It was concluded that as the temperature at which this fish is grown increases, the growth of the fish will increase. Which of the following assumptions must have been made when drawing this conclusion?

1) Fish growth can be decreased by decreasing temperature.
2) Temperature affects fish growth.
3) Light has no effect on fish growth.
4) All other conditions were identical among the experimental fish.
5) All fish can grow at 25° C.

17. A biologist conducted an investigation on the effects of pH on algal growth. The following data were collected.

pH	# of algal cells/ml sampled
4	200
5	400
6	600
7	800

These data were then presented to a group of students who concluded that an increase in pH will result in an increase in algal growth. Which of the following assumptions must be made by the students?

1) Temperature does not affect algal growth.
2) All other variables have been controlled.
3) There is enough light to support cell growth.
4) There were the same number of cells at each pH in the beginning of the experiment.
5) All of the above assumptions must be made.

18. A biologist noticed some ornamental plants growing in a yard that he passed while walking to work. At the other edge of the yard, the plants were eighteen inches tall. The farther away from the edge of the yard, the shorter the plants were until, at the other end of the yard where a spruce tree grew, they were only three inches tall. The biologist's interpretation was that the spruce tree deposited some chemical in the soil that inhibited the growth of the plants. Which of the following assumptions did he most likely make?

1) Spruce juice is soluble in water.
2) Growth of the ornamental plants is inhibited by spruce trees.

3) The ornamental plants were all the same size when they were planted.
4) Growth of the ornamental plants is stimulated by calcium carbonate leaching out of the concrete sidewalk by the edge of the yard.
5) The yard slopes from front to back.

19. Some students added 1000 cells to a flask of nutrient medium and placed them in an incubator. The next day the flask contained 3000 cells. The students suggested the following interpretations.

    A. Every cell divided at least once.
    B. Some cells did not divide, but others divided more than once (i.e., some of the daughter cells produced by each of the original parent cells also divided).
    C. Every cell divided more than once (i.e., all daughter cells also divided).

Which of these would be acceptable interpretations?

1) A only.
2) A and B only.
3) B and C only.
4) B only.
5) A, B, and C.

20. Your friendly neighborhood farmer has a flock of chickens that roam freely in the barnyard. One nice sunny warm afternoon, when things are peaceful, you notice that many of the birds sit down on bare patches of dry soil and ruffle their feathers so vigorously that clouds of dust are tossed up and over the birds. You recognize the behavior pattern as "dustbathing" and conclude that this is the way the birds keep clean, i.e., get rid of ectoparasites such as lice and feather mites on their bodies. Which of the following assumptions did you make in reaching this conclusion?

1) Dustbathing is a genetically controlled behavior.
2) The birds are normally infested with ectoparasites.
3) Birds tend to their body care needs only when things are peaceful.
4) Dustbathing could also serve other functions, such as nest-cup digging.
5) This behavior occurs only on warm sunny days when ectoparasites are likely to be active.

21. The farmer is indignant: he claims that his flock is clean, that they are not infested with ectoparasites. To prove his point, he captures 10 birds, 5 of which had not dustbathed and 5 of which had just finished dustbathing. None of the 10 birds show any signs of being infested with ectoparasites. You challenge his proof on the basis that:

1) The five that did not dustbathe could have already been clean and therefore did not need to perform the behavior; the 5 that had just finished could have been cleaned by dustbathing.
2) Ten birds are not enough of a sample on which to base valid conclusions.
3) Whether or not the birds are presently infested with ectoparasites is immaterial; what is important is that some time in the distant past they must have been.
4) It is a well known fact, and repeated in many poultry raising books, that dustbathing serves to control ectoparasites on birds.
5) He would of course be biased since he has a vested interest in convincing you that his birds are clean.

22. You decide to pursue the matter further by conducting your own experiment. Two identically clean areas are prepared, each containing a surplus of food, water and dustbathing sites. Ten birds (Group A) that are loaded with ectoparasites are placed in one area, and ten

ectoparasite-free birds (Group B) are placed in the other. According to your hypothesis (that dustbathing is solely a response to ectoparasite infestations) you would predict that:

1) Group A birds should dustbathe throughout the day while Group B birds dustbathe only in the morning.
2) Group A birds should dustbathe more than do Group B birds.
3) Group A birds should dustbathe only in the morning.
4) Group A birds should dustbathe far less often than do Group B birds.
5) Only Group A birds will dustbathe.

23. The data reveal that the birds in both groups dusted about the same amount, and that all dustbathing occurred in the middle of the afternoon. You reject your original hypothesis (ectoparasites) and now decide to examine other variables that might be related to the expression of dustbathing. Which of the following variables is most likely to be involved?

1) The amount of noise from passersby and traffic.
2) The time at which the sun rises in the morning.
3) The number of birds in the flock.
4) The location and amount of food available to the birds.
5) The amount of time that you have to make observations.

24. A biologist investigated plants' bending response to light. He made the hypothesis that the plant tip somehow senses the direction of the light source. He covered the tips of growing plants with a black cap. These plants did not bend toward a light source, but plants with tips uncovered did bend. He concluded that these results supported his hypothesis. Suppose he repeated this experiment 10 times and every time observed the same results. These repetitions:

1) Show he is a good experimenter.
2) Increase his confidence in his interpretation.
3) Increase his confidence in his experimental design.
4) Increase his confidence in his results.
5) Increase his confidence in his hypothesis.

25. Which of the following assumptions was the scientist in the previous question most likely making?

1) The caps were allowing some light to reach the tips.
2) The weight of a cap on the tip did not prevent bending.
3) Plants can bend toward a light source when their tips are not covered.
4) The tip of the plant can sense the direction of a light source.
5) The scientist was making no assumptions.

26. The biologist in the previous two questions later decided that he had not used adequate controls in his experiment. Which of the following would be the best control?

1) Grow plants in complete darkness and see if they bend.
2) Cut off tips of the plants and see if they bend toward a light.
3) Place a transparent cap on the tip and see if the plant bends toward light.
4) Grow plants lighted from all directions and see if they bend.
5) The control he used, plants without their tips covered, could not be improved.

27. Humans have a relatively high intake of sugars, and although the human body can synthesize starch, it cannot synthesize cellulose. From this information which of the following would be the best conclusion?

1) Humans do not have the proper enzymes to digest or break down sugars.

2) Humans do not have the proper enzymes to polymerize sugars.
3) Cellulose and starch are both polymerized from sugars using the same enzyme.
4) Enzymes which synthesize cellulose are the same enzymes which break it down.
5) Humans do not have the proper enzymes to synthesize cellulose.

28. A student suggested that the rate of a reaction would be reduced if molecules of an inert substance were added to the mixture containing the reactive molecules. He said that collisions would occur between reactive and inert molecules but would not lead to reactions. Which of the following assumptions was he making?

1) Adding the inert molecules raises the temperature of the mixture.
2) There are fewer collisions between reactive molecules after the inert molecules are added.
3) The inert molecules change the orientation of reactive molecules so that all molecular collisions occur the same way.
4) Reactive molecules bounce off inert molecules at a greater velocity than they had before the collision.
5) Collisions with inert molecules result in breakage of newly formed chemical bonds and reformation of reactant molecules.

29. A biochemist was studying the combination of two reactants (A and B) to form a single product (C). An enzyme catalyzed the reaction. After the reaction had started, the biochemist added another chemical (X) which rapidly combined with A, preventing any reaction with B. She added an amount of X sufficient to combine with about ½ the reactant A. Which of the following would most likely occur after X was added?

1) The reaction of A and B to produce C would proceed more rapidly.
2) The reaction of A and B to produce C would proceed at the same rate.
3) The reaction would stop (i.e., no more C would be produced).
4) The reaction of A and B to produce C would proceed more slowly.
5) Contact between the enzyme molecules and any A molecules would be prevented.

30. The biochemist in the previous problem repeated her experiment at a higher temperature. All other conditions and procedures were identical to those in the first experiment. In this second experiment which of the following would most likely occur after X is added?

1) The reaction of A and B to produce C would proceed more rapidly than before X was added.
2) The reaction of A and B to produce C would proceed at the same rate as it did before X was added.
3) The reaction would stop (i.e., no more C would be produced).
4) The reaction of A and B to produce C would proceed more slowly than before X was added.
5) Contact between the enzyme molecules and any A molecules would be prevented.

31. The student in the previous problem was in a hurry to finish her project by the end of the laboratory period. Since the enzyme was very expensive, she decided to maintain a low enzyme concentration but incubate at a higher temperature. Which of the following was she most likely assuming?

1) Temperature does not affect reaction rate.
2) The higher temperature alters the enzyme so that it is a better catalyst.
3) The higher temperature does not alter the enzyme's structure.
4) The reaction rate slows at higher temperatures.
5) Enzymes are proteins, polymers of amino acids.

32. The function of an enzyme can sometimes be stopped by adding a "poison" in much smaller concentrations than the reactants. Which of the following is the best hypothesis regarding the mechanism of the "poison's" reaction?

    1) It combines permanently with the enzyme at the site where reactants normally combine.
    2) It causes the enzyme to release reactants more quickly after they have combined.
    3) It speeds up the molecular motion of the reactants and enzyme so they are moving more rapidly. .
    4) It combines permanently with reactants so they can not combine with each other.
    5) It causes the enzyme and reactants to collide with a greater force.

33. Some students added an enzyme to a flask containing a pure polymer sample. Several hours later they discovered that the flask contained a significant amount of amino acid. Student A concluded that the polymer was a protein. Student B concluded that the enzyme had broken down to monomers. The data are consistent with the conclusion of:

    1) A only
    2) B only
    3) A and B
    4) Neither A nor B
    5) The data are insufficient to decide which of the above four choices is best.

34. The students in the previous question waited several more hours and again measured the contents of the flask. If Student A's conclusion is valid, which of the following would they most likely find?

    A. Less amino acid.
    B. More amino acid.
    C. More enzyme.
    D. Less enzyme.

    1) B and C
    2) B and D
    3) A and C
    4) C only
    5) B only

35. The students in the previous question now repeated their first experiment five times. Each time they obtained results identical to those of the original experiment. These additional results support the conclusion of:

    1) A only
    2) B only
    3) A and B
    4) Neither A nor B
    5) The data are insufficient to decide which of the above four choices is best.

36. A student homogenized some potato tuber tissue. After analyzing the resulting mixture she found that it contained 12 gram/liter of starch. She then added two drops of a protein solution and measured the amount of sugar present in the mixture. At various times after adding the protein she obtained the following data:

Time (hrs)	0.50	1.00	1.50	2.00
Sugar Content (g/l)	0.08	0.13	0.17	0.24

Which of the following would you also expect to find?

1) An increase in starch content as time increased.
2) No sugar present if measured at time = 0.
3) An increase in protein as time increased.
4) A decrease in starch content as time increased.
5) Both 1 and 3.

37. Suppose the student in the previous question now cooled the mixture 10 °C after the 2 hour measurement. After another ½ hour she made another measurement of the sugar content. Which of the following values would she most likely find?

1) 0.08 g/l
2) 0.13 g/l
3) 0.24 g/l
4) 0.26 g/l
5) From the information given, no prediction is possible.

38. A student had a solution which contained only water and polymers extracted from cells. He predicted that after standing for two hours the solution would contain nucleotides and sugars. Which of the following assumptions did he most likely make?

The original solution contained:

1) Nucleic acid and polysaccharide.
2) Protein and polysaccharide.
3) Protein and nucleic acid.
4) Nucleic acid, protein and polysaccharide.
5) Lipid, nucleic acid, protein and polysaccharide.

39. A cell biologist was examining sections of adrenal gland under the light microscope. The cells had a "foamy" appearance because they contained very many small, round, clear spots that looked like tiny bubbles. The biologist made three hypotheses:

A. The spots were artifacts, caused by destruction of the cell material during processing for microscopy.
B. The spots were empty bubbles produced by normal life processes, which had some important function in the cell.
C. The spots were really filled with cellular material, but the material did not react with the stain he used.

Which of these hypotheses adequately account for the observation?

1) A, B, and C
2) A and B only
3) A and C only
4) B and C only
5) A only

40. The cell biologist in the previous question examined some adrenal gland cells in the electron microscope. The cells were filled with round, clear spots but with the greater resolution in this kind of microscope it was possible to see that each spot was surrounded by a membrane that was identical in appearance to the membrane around the outside of the cell. Which of the original hypotheses adequately account for all the observations?

1) A, B, and C
2) A and B only
3) A and C only
4) B and C only

5) A only

41. A certain chemical is known to react with sugars to produce a red-colored product. It is used to detect sugars in tissue sections prepared for examination under the microscope. Which of the following would be most likely to produce artifacts that could lead to a false conclusion?

1) The chemical reacts with sugars which are polymerized.
2) The chemical reacts with sugar monomers.
3) The color can be bleached by treatment with another chemical.
4) Sugar monomers dissolve in the chemical solution before reacting.
5) Sugar polymers do not dissolve in the chemical solution but react as solids.

42. An entomologist was studying some microscope slides of the salivary gland of the fruit fly. When preparing the slides he had applied a stain known to indicate the presence of nucleic acids. His observations indicated heavy staining of the nuclei of the salivary gland cell but all other structures appeared to be unstained. Which of the following would be the best conclusion you could draw from these data?

1) Nucleic acid is a polymer.
2) Polymers are found in all parts of the cell.
3) The nuclei contain at least one polymer.
4) Salivary gland cells contain more nucleic acid than most other cells.
5) Nucleic acid is composed of monomers.

43. A biologist was studying the cells of a certain organ. He embedded a piece of the tissue, cut sections and stained the resulting slides with a single stain. The thickness of each section was one twentieth the thickness of an average cell. He examined the cross section of seven cells on one slide. In one cell he saw a small round structure which was the same color as the stain he had applied. When he looked at unstained cells he saw no such structure. Which of the following is his best interpretation?

1) The structure is an artifact.
2) The structure is real and occurs in only a few cells.
3) The structure is real, occurs in all cells but is missed by the plane of the section in most cells.
4) Both 2 and 3 are equally valid interpretations.
5) 1, 2, and 3 are all equally valid interpretations.

44. Next the biologist examined a total of seven sections, each containing the cross sections of seven cells. In each section he saw the small round structure in one cell. Now which of the following is his best interpretation?

1) The structure is an artifact.
2) The structure is real and occurs in only a few cells.
3) The structure is real, occurs in all cells but is missed by the plane of the section in most cells.
4) Both 2 and 3 are equally valid interpretations.
5) 1, 2, and 3 are all equally valid interpretations.

45. Using other pieces of the same kind of tissue he prepared seven new slides. For each slide he employed a different method of processing the tissue and a different stain. The thickness of all sections remained the same. In each of these sections, one cell of seven contained a small round structure colored the same as the staining solution. Which of the following is his best interpretation?

1) The structure is an artifact.

2) The structure is real and occurs in only a few cells.
3) The structure is real, occurs in all cells but is missed by the plane of the section in most cells.
4) Both 2 and 3 are equally valid interpretations.
5) 1, 2, and 3 are all equally valid interpretations.

46. The biologist decided to study this problem by using cell fractionation. He homogenized a piece of the tissue and separated four fractions by centrifugation. One fraction consisted entirely of small round structures which could be colored by each of the stains he had used on his slides. Which of the following is his best interpretation?

1) The structure is an artifact.
2) The structure is real and occurs in only a few cells.
3) The structure is real and occurs in all cells but is missed by the plane of the section in most cells.
4) Both 2 and 3 are equally valid.
5) 1, 2, and 3 are all equally valid.

47. Finally he identified a certain protein in the tissue. One gram of tissue contained 25 mg of this protein. He homogenized one gram of tissue and separated the fractions by centrifugation. The fraction which contained the small round structures also contained all 25 mg of the certain protein. Which of the following is his best interpretation using all the data?

1) The structure is an artifact.
2) The structure is real and occurs in only a few cells.
3) The structure is real, occurs in all cells but is missed by the plane of the section in most cells.
4) Both 2 and 3 are equally valid interpretations.
5) 1, 2, and 3 are all equally valid interpretations.

48. Two scientists (A and B) homogenized a sample of muscle tissue. Subsequent measurements indicated a high sugar content in the homogenate. Scientist A hypothesized that the starch content of the original muscle tissue was high but the sugar content was low. Scientist B hypothesized the opposite, that the sugar content of the original muscle tissue was high. Which of the following was Scientist A most likely assuming?

1) Sugar is necessary for muscle tissue to function.
2) The starch content of the original muscle tissue was high.
3) Homogenization can indicate the presence of starch.
4) Homogenization can break down starch into monomers.
5) Both 2 and 3.

49. Additional measurements showed that the sugar content of the original muscle tissue in the above question was low. Scientist B now changed his hypothesis to: homogenization releases or activates enzymes which can break down starch into monomers. Which of the following would most likely test this hypothesis?

1) Homogenize the tissue for a longer time and determine whether more sugar is present.
2) Add starch to homogenized muscle tissue and measure starch concentration at later times.
3) Measure the protein content of the original muscle tissue and homogenized muscle tissue.
4) Add an enzyme (known to break down starch) to homogenized muscle tissue and measure starch concentration at later times.
5) All the above would be equally valid tests of the hypothesis.

CELL AND LIFE CYCLES

Mitotic cell division is characterized by well-defined changes and movements of the chromosomes. Such movements are used to identify stages: prophase, metaphase, anaphase, telophase. During anaphase the two identical chromatids, which make up each chromosome, separate and move to opposite ends of the cell which then divides. The resulting two daughter cells are identical to each other and to the parent cell and contain equal numbers of chromosomes. The number and structure of these chromosomes are revealed by a karyotype. Alterations in the mitotic process may result in unequal numbers of chromosomes.

Organisms may reproduce asexually by mitosis only, but those reproducing sexually exhibit a second type of cell division. At some stage of sexual reproduction meiosis occurs, usually in small, localized and highly specialized portions of an organism's body. Complete meiotic division encompasses two cell divisions so that a single parental cell produces four daughter cells. The first meiotic division is characterized by the pairing of homologous chromosomes (synapsis). Daughter cells of the first division contain only one chromosome from each homologous pair and thus one-half the chromosome number of the parental cell. The second meiotic division is essentially identical to mitosis. As a consequence, cells with a haploid number of chromosomes are formed from the original parent cell with a diploid number of chromosomes.

In most animals and some plants which reproduce sexually meiosis produces gametes (male or female sex cells) directly. A male gamete (sperm) and female gamete (egg) fuse at fertilization, producing a zygote, which usually divides repeatedly by mitosis and ultimately produces a mature individual.

In some sexually reproducing organisms meiosis produces reproductive spores. A spore may divide mitotically to produce a multicellular body and subsequently to form gametes. The specific pattern of mitosis and meiosis to form gametes, spores, and multicellular bodies is termed the life cycle.

EXAMPLE 1

The diploid number of chromosomes from a certain animal is twelve. You observe a cell which contains two distinct and separate groups of six chromosomes. This cell is:

1) In metaphase of mitosis.
2) In anaphase of mitosis.
3) In metaphase of the first division of meiosis.
4) In anaphase of the second division of meiosis.
5) Between meiotic divisions.

ANALYSIS:

Since the cell contains two distinct and separate groups of chromosomes it is probably in anaphase or telophase of a division process. In metaphase all the chromosomes will be in one group. Since it is from an animal, it must have begun the division process as a diploid cell. The only haploid cells in animals are gametes which do not divide. If a diploid cell with twelve chromosomes undergoes mitosis, each chromosome will separate into two and there will be two distinct groups of twelve chromosomes at anaphase. In the second meiotic division, the cell will begin with six chromosomes, each will separate into two, and two separate groups of six will form at anaphase. Thus 4 is the best choice.

EXAMPLE 2.

A biologist measured the amount of DNA in several single diploid cells taken from a culture. She knew that these cells could divide mitotically. Her data are shown below:

Cell	1	2	3	4	5	6	7
DNA Content	11.1	5.4	7.3	10.9	5.6	8.6	11.0

Which of these cells have most likely just completed mitosis?

1) 1, 4 and 7
2) 3 and 6
3) 2, 3, 5 and 6
4) 2 and 5
5) No conclusion can be made from the data above.

ANALYSIS:

The minimum amount of DNA/cell (1X) should be present in cells immediately after mitosis is complete. DNA/cell will remain constant until a cell enters S phase, in which DNA is replicated. It will increase throughout S phase until replication is complete and the DNA/cell (2X) is double the previous amount. At mitosis half the DNA will be distributed to each daughter cell and DNA/cell will again equal 1X. Thus, cells 2 and 5 have just completed mitosis, cells 3 and 6 are in S phase, and cells 1, 4 and 7 are between the completion of S phase and the beginning of mitosis.

1. Which of the following could NOT be detected by examining a karyotype?

    1) The way chromosomes move during division.
    2) The addition of a chromosome.
    3) The loss of a chromosome.
    4) Breakage of a chromosome into two parts.
    5) Either 2 or 3.

2. Suppose you are studying a newly discovered plant. You remove one cell from it and treat it with a chemical which makes it stop mitosis at metaphase. You preserve and stain the treated cell and prepare a karyotype from it. There are 15 chromosomes. How would you interpret this observation?

    1) This cell may have gained a pair of chromosomes.
    2) This could be the normal chromosome number for this cell.
    3) When this cell was produced by mitosis, both halves of one chromosome might have gone to the same daughter cell.
    4) When you prepared the karyotype, one chromosome might have been lying directly above another of similar size and shape.
    5) All of the above.

3. Consider that in a culture of cells which normally have 46 chromosomes you find an occasional cell with 92 chromosomes. Assume that the cause for the increase in number is traceable to the mitotic process. Which one of the following best explains what might have occurred?

    1) The chromosomes failed to shorten and thicken.
    2) The chromosomes failed to separate into two and migrate to each end of the cell.
    3) The chromosomes separated into two but did not migrate to opposite ends of the cell.
    4) The cell divided with unequal numbers of chromosomes.
    5) The chromosomes did not align themselves in the middle of the cell.

4. A cell with 14 chromosomes undergoes mitosis and one of the daughter cells ends up with 13 chromosomes and the other with 15. At which point in mitosis would you suspect that something went wrong?

    1) Prophase
    2) Anaphase
    3) Metaphase
    4) Telophase
    5) 1 and 4 above.

5. In order to avoid the doubling of chromosomes at every generation, meiosis produces sperm and eggs with fewer chromosomes than the parent cell has. Which of the following is the strongest argument against the prediction that meiosis will produce daughter cells with unequal numbers of chromosomes?

    1) All division processes in nature produce equal parts.
    2) Chromosomes split down the middle in cell division.
    3) Chromosomes come in pairs.
    4) Every sperm cell has an equal opportunity to fertilize every egg cell.
    5) Only one sperm cell can fertilize an egg.

6. While examining the karyotype of an organism, a biologist noticed that part of one chromosome had broken off and was not present.

Suppose this cell now produced gametes by meiosis. In what percentage of the gametes would the broken chromosome be found?

1) 0%
2) 25%
3) 50%
4) 75%
5) 100%

7. The cells in the outermost layer of the seminiferous tubules in the testis of a certain species of mammal have 12 double-stranded chromosomes. Adjacent to these cells is a layer of cells, called secondary spermatocytes, produced by the first meiotic division of the outermost cells. The secondary spermatocytes of this mammal should have:

1) 12 double-stranded chromosomes.
2) 12 single-stranded chromosomes.
3) 6 double-stranded chromosomes.
4) 3 double-stranded chromosomes.
5) 6 single-stranded chromosomes.

8. After observing chromosomes in a single cell from a multicellular organism, a student stated that this cell could be either haploid or diploid. Which of the following must the student have observed?

1) Chromosomes in a gamete.
2) Chromosomes at synapsis.
3) Chromosomes with no homologues present in the cell.
4) An even chromosome number.
5) Either 1 or 4.

9. Upon microscopic examination of cells from a small organism a student found a cell in which it appeared that similar chromosomes were paired along a plane in the central region of the cell. From what the student learned in Biology, he would most likely conclude that the cell was undergoing:

1) Mitotic division (mitosis).
2) A first meiotic division (meiosis I).
3) A second meiotic division (meiosis II).
4) Either mitosis or meiosis I.
5) Either mitosis or meiosis II.

10. A certain kind of cell could undergo either meiosis or mitosis. You are asked to determine which process is occurring. A microscope has been set up for your observations of the living cell. Which of the following observations would allow you to most confidently conclude that meiosis was occurring?

1) You observed homologous pairs of chromosomes in the cell before division.
2) You observed the haploid number of chromosomes in the cell before division.
3) You observed the haploid number of chromosomes in the daughter cells.
4) You observed the diploid number of chromosomes in the cell before division.
5) Both 3 and 4 together are necessary.

11. Each chromosome bears a single centromere. The centromere is a structure which in the double-standed chromosome holds the two strands of the chromosome together and which is itself divided when the two strands separate. A biologist observes a cell. The chromosomes are distinct and double-stranded and the biologist can accurately count the number of centromeres. A little while later he observes the SAME cell. The chromosomes are now in two bunches on opposite sides of the cell, but the number of centromeres in the cell has not changed and there has been no division of the cytoplasm. The best conclusion is:

1) The cell has gone from prophase to anaphase of mitosis.
2) The cell has gone from prophase to anaphase of the first meiotic division.
3) The cell has gone from prophase to anaphase of the second meiotic division.
4) The cell has undergone fertilization.
5) The cells were NOT undergoing mitosis, because the chromosomes in mitosis are single-stranded.

12. A student observed a cell which contained 4 chromosomes, all of different sizes and shapes. All 4 chromosomes had the typical double-stranded appearance. Which of the following would be possible statements about this cell?

1) It is a zygote.
2) It has been produced by mitosis.
3) It has completed meiosis I but not meiosis II.
4) Both 2 and 3.
5) All the above.

13. A plant is found to have 14 chromosomes. Another plant which looks almost identical is found to have 21 chromosomes. A biologist concluded that these plants could have come from the same parent plant because of a failure of some meiotic cells to divide after meiosis I. Which of the following is he assuming?

1) Meiosis results in equal numbers of chromosomes in gametes.
2) Reproduction was asexual in these plants.
3) Meiosis in these plants must result in the haploid number of chromosomes.
4) Gametes that contain a diploid number of chromosomes could fertilize gametes that contain a haploid number of chromosomes.
5) The haploid number of chromosomes of this plant is 21.

14. A student concluded that all 4 cells produced from one cell by meiosis were genetically identical. Which of the following was this student most likely assuming?

1) In the original cell, replication of the chromosomes occurred after the first meiotic cell division.
2) In the original cell, both chromosomes in every homologous pair were genetically identical.
3) Genetically identical cells will have the diploid number of chromosomes.
4) In the original cell, replication of the chromosomes occurred after the second meiotic cell division.
5) Both 1 and 3 are necessary assumptions for the student to draw her conclusions.

15.   A biologist marked a cell which she knew was about to undergo meiosis. A short while later she observed the 4 cells produced by the original marked cell; their chromosome numbers were: 17, 17, 18, 16. She knew these numbers indicated that something abnormal had occurred during meiosis. Which of the following most likely occurred?

1) One chromosome did not split apart in the first meiotic cell division.
2) Synapsis did not occur properly in the second meiotic cell division but other events occurred as usual.
3) One homologous pair of chromosomes did not separate in the first meiotic cell division but moved to one cell and later separated.
4) In the second meiotic division one chromosome did not split apart but moved to one cell and later split.
5) None of the above could account for these chromosome numbers.

16.   The biologist in Question 15 had not observed the chromosome number in the original marked cell. Which of the following was most likely the chromosome number of that cell?

1) 33
2) 34
3) 35
4) 36
5) 68

17.   A biologist applied chemical M-P to 1000 cells about to undergo meiosis. When he examined the chromosome numbers of the resulting gametes he found that half had 8 chromosomes and the other half had 10 chromosomes. Which of the following hypotheses is consistent with his observations?

1) Chemical M-P actually had no affect since chromosome separation is random at meiosis.
2) Chemical M-P prevented separation of one homologous pair at meiosis I.
3) Chemical M-P prevented "splitting" of one chromosome at meiosis II but allowed "splitting" at gamete formation.
4) Chemical M-P caused inversion of homologous pairs at synapsis.
5) Both 2 and 3.

18.   In question 17, if chemical M-P was NOT added what would be the expected gamete chromosome number and the resulting zygote chromosome number from the fertilization of such gametes?

1) 11 and 22
2) 9 and 18
3) 8 and 16
4) 7 and 14
5) cannot tell

19.   A physiologist treated some newborn hamsters with a substance which prevented any meiotic cell divisions in the animals for the rest of their lives. Which of the following would be the best prediction regarding these treated animals?

1) Their growth and mature size will not be significantly affected.
2) Their size increase will be about 1/2 that of untreated hamsters.
3) The animals will exhibit no size increase after the chemical is applied.
4) There will be a small effect on the animals occurring in all parts of the body.
5) The animals will probably die shortly after the substance is applied.

20.   A biologist studying development in ferns successfully fertilized an egg with two sperm cells. Later study of the zygote formed in this way showed that it contained 27 chromosomes. How many chromosomes were contained in the cells of the stems of the plant that produced the egg?

1) 9
2) 18
3) 27
4) 36
5) The number cannot be calculated from the observations given.

21.   A student mixes eggs and sperm from a certain fish (2N = 12). He finds that 5 of the resulting zygotes contain the following numbers of chromosomes: 11, 16, 19, 23, and 20. He states that these chromosome numbers could not have resulted from multiple fertilization alone. Which of the following assumptions is he making?

1) Eggs and sperm all carry the haploid number of chromosomes.
2) Some chromosomes are accidentally lost when multiple fertilization occurs.
3) Egg cells possess mechanisms to prevent multiple fertilization.
4) Chromosomes come in pairs.
5) Homologous chromosomes are separated during mitosis.

22.   Suppose you are studying some cells from a chick embryo. You find that about half the cells contain twice as much DNA as others. Which of the following is the best explanation?

1) The cells with low DNA content are gametes.
2) The cells with high DNA content are zygotes.
3) The cells with high DNA content resulted from abnormal mitosis.
4) The cells with low DNA content contain broken chromosomes.
5) These observations are the expected result.

23.   A student measured the DNA content per cell during the meiotic process at the following points:

   A.   Just before the first meiotic division
   B.   Just after the first meiotic division but before the second meiotic cell division.
   C.   Just after the second meiotic cell division.

Which of the following values did she most likely obtain?

1) A=30, B=15, C=15
2) A=40, B=20, C=10
3) A=10, B=10, C=5
4) A=10, B=5, C=5
5) A=40, B=20, C=30

24.   In some green algae gametes can be produced by meiosis or mitosis. If the spore contains 12 chromosomes and the zygote 24 how many chromosomes would the gametes produced by mitosis have?

1) 12
2) 24
3) 36
4) 6
5) None of the above

25.   The fern is a plant in which the life cycle includes mitosis, meiosis and fertilization. Which of the following is the best statement about the fern life cycle?

1) Meiosis produces gametes and fertilization produces spores.
2) Mitosis produces gametes and fertilization produces a zygote.
3) Meiosis produces gametes and fertilization produces a zygote.
4) Meiosis produces gametes and mitosis produces spores.
5) Meiosis produces spores and fertilization produces gametes.

26. In studying an unknown organism, a scientist finds that at one point in its sexual life cycle, two of its cells fuse together, starting a new generation. Which of the following could he then conclude?

    1) That meiosis must also be present in its life cycle, in order to go from 2N to 1N.
    2) That the organism is multicellular.
    3) That the organism is an animal, since it has only one growth and development phase.
    4) That the organism will have no asexual reproduction, since it can reproduce sexually.
    5) That all the above are true.

27. Sexual reproduction in ploimate rotifers is characterized by production of eggs which, if fertilized, develop into females. Eggs which are not fertilized develop into males. Which of the following statements is NOT acceptable?

    1) The unfertilized eggs are haploid.
    2) The males are haploid.
    3) The females are haploid.
    4) The females are diploid.
    5) The zygotes are diploid.

28. In some organisms, such as the green alga *Ulothrix*, the zygote is the only diploid cell in the life cycle, all others are haploid. In organisms of this type, which of the following must also be so?

    A. Gametes are produced by meiotic cell division
    B. The zygote divides meiotically.
    C. Gametes are diploid.

    1) A only.
    2) B only.
    3) C only.
    4) A and B only.
    5) A, B, and C.

29. *Ulothrix* (see Question 28) can also reproduce asexually. One of the cells in a multicellular stage of the life cycle divides to produce free swimming cells called zoospores which germinate and grow into another multicellular body. In asexual reproduction in this plant which of the following are most likely?

    A. Zoospores are haploid.
    B. Zoospores are produced by meiotic cell division.
    C. Mitotic cell division occurs during growth.

    1) A only.
    2) B only.
    3) C only.
    4) A and C only.
    5) A, B, and C.

30. A botanist was studying a colony of small, green, leafy plants growing in a damp area. He knew that these plants reproduced sexually. When examining cells from the leafy portion of several plants, he found that all cells contained 83 chromosomes. Which of the following would be the most likely statement to make about these plants?

    1) The diploid number for this organism is 83.
    2) The haploid number for this organism is 164.
    3) The haploid number for this organism is either 42 or 41.
    4) The haploid number for this organism is 83.
    5) Both 1 and 2 are equally likely as the best choice.

31. The botanist in Question 30 later discovered that each of the green leafy plants grew from a single-celled structure which apparently had fallen on the ground and germinated. This single-celled structure is most likely which of the following?

    1) An unfertilized female gamete
    2) A spore
    3) A male gamete
    4) A zygote
    5) A diploid gamete

32. If the botanist in Questions 30 and 31 made a detailed study of the complete life cycle of the organism, how many homologous pairs of chromosomes would he most likely find in a diploid cell?

    1) 83
    2) 164
    3) Either 41 or 42
    4) All the above are equally likely.
    5) Cannot be determined from the information given.

The following information applies to the next three questions.

In flowering plants pollen grains are produced by meiosis. Shortly after reaching the female part of the flower the pollen grain germinates and a pollen tube grows through the female structure to the egg cell. After germination, the single nucleus in the pollen grain divides and one of these nuclei again divides resulting in 3 nuclei in the pollen tube. One of these nuclei will fuse with the egg cell to form a zygote.

33. In a plant which has a diploid chromosome number of 34, what will be the chromosome numbers of the 3 nuclei in the pollen tube?

    1) 34, 34, 34
    2) 34, 34, 17
    3) 34, 17, 17
    4) 17, 17, 17
    5) None of the above could be the correct numbers.

34. What will be the chromosome number of the zygote?

    1) 34
    2) 6
    3) 51
    4) 17
    5) It is not possible to deduce the chromosome number in this organism.

35.   Immediately after fertilization a cell is found in the ovule (the structure containing the fertilized egg) which has 51 chromosomes. This cell is:

1)   The zygote.
2)   One of the remaining nuclei from the pollen tube.
3)   The product of the fusion of 3 haploid cells.
4)   One which will divide mitotically to produce the new organism.
5)   Either choice 1 or 4.

MENDELIAN GENETICS

Characteristics or traits, such as eye and hair color, are transmitted from parents to offspring. Traits which are observed in an individual are termed the phenotype, although genetic information for other traits may be carried without being expressed. The total genetic information carried by an organism, both expressed and unexpressed, is referred to as the genotype.

Each trait is governed by a "factor", called a gene, which is part of a chromosome. Each gene may exist in alternative forms, alleles, which are expressed in the different forms that the trait may take. Diploid cells have two alleles governing every trait and the alleles exist at identical positions on two homologous chromosomes. If the alleles for a trait are identical, the cell and the individual of which it is a part are said to be homozygous. If the alleles are different, the cell is heterozygous.

If only one allele is visibly expressed in the phenotype of a heterozygous individual, that allele is called dominant. The allele which is not expressed is called recessive. Often the phenotype of the heterozygote is intermediate between those of the homozygotes or includes expression of both alleles. These conditions are called incomplete dominance or intermediate inheritance. Dominance and recessiveness are useful descriptive terms, but do not explain inheritance patterns. Satisfactory explanations require concepts from biochemistry and molecular biology.

Genes which are located on the same chromosome are said to be linked and are usually transmitted to offspring as a unit. In some cases, however, homologues may become entangled at synapsis and portions of the chromosomes may be exchanged (crossing-over).

In most organisms the sex of an individual is determined by sex chromosomes. In humans and many other higher organisms the male sex chromosomes are of different sizes, a small Y chromosome and a larger X chromosome. Females possess two X chromosomes. Both sex chromosomes carry genes for traits other than those associated with sexual characteristics. Since the X chromosomes are larger than the Y chromosomes, they carry alleles which are not paired on their smaller homologues. Such genes are termed sex-linked.

EXAMPLE 1

A pet store owner felt that if he could breed a special mouse he would increase his sales. From a mouse breeders' catalog he knew the following traits were dominant — red eyes, white hair, rough coat, short tail, while blue eyes, gray hair, smooth coat and long tail were recessive. From the same catalog he ordered a pair of mice which were heterozygous for red eyes and white hair while homozygous for rough coat and short tails. What percentage of the offspring from the cross will have blue eyes, gray hair, rough coat and long tail?

1) 75%
2) 45%
3) 33%
4) 25%
5) 0%

ANALYSIS:

The parent mice are homozygous for short tails, therefore all their gametes will carry the gene for short tail and all their offspring will be homozygous for this trait and have short tails. None of the offspring will have long tails unless a certain mutation occurs.

EXAMPLE 2

Suppose that in one family there were two children, a boy and a girl. You know that the girl is color blind but the boy has normal vision. Which of the following could you most likely conclude about the father and/or mother?

1) The mother has normal vision.
2) The father has normal vision.
3) Both the father and mother are color blind.
4) Neither the father nor mother are color blind.
5) None of the above could be concluded.

ANALYSIS:

Color-blindness is a recessive sex-linked trait. The color-blind daughter therefore is homozygous for the trait and received one X chromosome bearing the gene from each of her parents. The father, then, has one gene for color-blindness on his X chromosome and no genes governing color vision on his Y chromosome and so is color-blind. Since the son must have received his Y chromosome from his father, he must have received his gene for normal vision on the X chromosome from his mother. The mother therefore is heterozygous and has normal vision.

The drawing, which refers to Questions 1 and 2, represents a cell with its chromosomes. The letters indicate genes associated with the chromosomes.

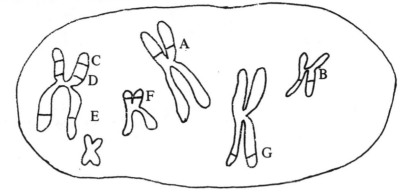

1. How many chromosomes would be present in each gamete if the above cell underwent meiosis?

   1) 3 double-stranded chromosomes
   2) 6 single-stranded chromosomes
   3) 3 single-stranded chromosomes
   4) 12 single-stranded chromosomes
   5) 12 double-stranded chromosomes

2. If the above cell underwent meiosis, which of the following genes would most likely be present in the same gamete?

   1) A and G
   2) C, D, and E
   3) B and F
   4) A, C, and D
   5) A through G

3. A student says, "Homologous chromosomes in a cell do not contain the same information because loss of a piece of only one homologue results in an altered cell." Which of the following assumptions is he making?

   1) Information from both homologues is expressed in a cell.
   2) DNA carries the genetic information.
   3) The chromosome was broken during anaphase.
   4) The chromosome was broken during meiosis.
   5) All of the information in a cell is not expressed.

4. The ability to taste phenylthiocarbamide is governed by a single pair of genes. A man and a woman who are both tasters have one child who is taster and then have a second child who is a non-taster. The most probable explanation for these observations is that:

   1) Both parents are heterozygous.
   2) One parent is heterozygous; the other parent is homozygous recessive.
   3) One parent is heterozygous; the other parent is homozygous dominant.
   4) Non-taster is dominant to taster; therefore the second child was a non-taster even though he also had a gene for taster.
   5) A mutation occurred in one of the parents during meiosis; the mutated gene was transmitted to the child who is a non-taster.

5.   Suppose a mouse heterozygous for fur color is crossed with an albino mouse. What percent of the offspring would be albino?

   1)   0%
   2)   10%
   3)   25%
   4)   50%
   5)   100%

6.   In a hypothetical organism fur color may be expressed as black or white and fur color is determined by a single pair of genes. If a heterozygous white female is mated with a black male, what percent of the offspring will be black?

   1)   100%
   2)   75%
   3)   50%
   4)   25%
   5)   0%

7.   Assume that you crossed two flowers, a red one and a white one, and all the hybrids were pink. Which of the following is the best prediction about the parents and the F2 produced by crossing the hybrids?

   1)   The parents are both homozygous, and the F2 would be 1:2:1, red:pink:white.
   2)   The parents are both homozygous, and the F2 would be 3:1, pink to white, because the F1 is heterozygous.
   3)   The parents must be heterozygous, or no difference would have shown up in the F1.
   4)   An obvious error has been made, because such a cross would not produce the results above.
   5)   The red parent is dominant, the white parent is recessive; the hybrid is pink because the red was diluted. The F2 will be all types, not of a particular ratio.

8.   Two black female mice were bred with the same brown male. In three litters female A produced 9 black offspring and 8 brown offspring, while in 3 litters female B produced 19 black offspring and no brown ones. Assuming that coat color is determined by a single pair of genes in mice, which of the following is the most likely set of parental genotypes?

   1)   The male is homozygous for brown, female A is heterozygous, and female B is homozygous for black.
   2)   The male is heterozygous, female A is heterozygous, and female B is homozygous for black.
   3)   The male is homozygous for brown, female A and female B are both heterozygous for black.
   4)   The male is heterozygous, and both females are homozygous for black.
   5)   All 3 parents are heterozygous.

9.   A student placed a white female guinea pig in a cage with 3 male guinea pigs (one white and 2 black). Later the female gave birth to a white offspring. Which of the following is the best choice concerning the father of the offspring? (Note: Black is dominant.)

   1)   The white male must be the father.
   2)   One of the black males must be the father.
   3)   The white male could be the father.
   4)   Either of the black males could be the father.
   5)   Choices 3 and 4 together are the best choice.

30

10. If the student in Question 9 assumed both black males were homozygous, which of the following would be the best choice concerning the father of the white offspring?

    1) The white male must be the father.
    2) One of the black males must be the father.
    3) The white male could be the father.
    4) Either of the black males could be the father.
    5) Choices 3 and 4 together are the best choice.

11. An individual whose genotype is CCDDEE is crossed with one whose genotype is ccddee. C and D are on the same chromosome. D and E are on different chromosomes. Which kind (or kinds) of gamete will be LEAST often produced by the offspring?

    1) CDE
    2) cDE
    3) cdE
    4) cde
    5) cdE and CDe will be produced less often than CDE or cDE.

12. In guinea pigs eye color and fur texture are located on two separate chromosome pairs. In addition, brown eye color is dominant to pink eye color and rough fur is dominant to smooth fur. If a heterozygous brown-eyed guinea pig with a smooth coat (Bbrr) was mated to a pink-eyed guinea pig with a rough coat (bbRr) what percentage of the offspring would be heterozygous brown-eyed and smooth?

    1) 0%
    2) 18.75%
    3) 25%
    4) 56.25%
    5) 100%

13. In peas the genes affecting the shape and the color of the peas are located on two separate pairs of homologous chromosomes. If a plant heterozygous for yellow color and homozygous for round seed shape was crossed with a plant homozygous for green color and heterozygous for wrinkled seed shape, what percentage of the offspring would be heterozygous for yellow color and homozygous for round seed shape?

    1) 0%
    2) 18.75%
    3) 25%
    4) 56.25%
    5) 100%

14. Suppose you cross corn which is homozygous tall and heterozygous green-albino with corn which is heterozygous tall-short and homozygous green. What percentage of the offspring would be homozygous for at least one trait?

    1) 0%
    2) 25%
    3) 50%
    4) 75%
    5) 100%

15. In the cross described above (Question 14) what proportion of the progeny would be homozygous for both traits?

1) 0%
2) 25%
3) 50%
4) 75%
5) 100%

16. Suppose you cross corn which is homozygous tall and heterozygous green-albino with corn which is heterozygous tall-short and homozygous green. Just before crossing these plants, you spray them with a substance which would destroy any gametes that contain the genes for tall and for albino together. What percent of the offspring would be heterozygous tall-short and homozygous green?

1) 100%
2) 75%
3) 50%
4) 25%
5) 0%

17. In corn, smooth kernels are dominant over wrinkled and dark kernels are dominant over light kernels. A cross is made between a dark-smooth plant and light-wrinkled one. All offspring are dark-smooth. If these dark-smooth offspring are crossed among themselves, which of the following would be the most likely phenotype among their offspring?

1) Wrinkled-light.
2) Wrinkled-dark.
3) Smooth-light.
4) Smooth-dark.
5) There will be equal numbers of each of the above.

18. A plant geneticist crossed a tall sweet pea plant bearing green pea pods to a short sweet pea plant bearing yellow pods. All the offspring were tall plants bearing green pods. The geneticist hypothesized that the genes for height and pod color in this plant were linked. He crossed two of the offspring from the 1st generation and obtained the following offspring in the second generation.

Expression:	Tall Green	Tall Yellow	Short Green	Short Yellow
Number of Offspring:	34	11	13	6

If no crossing-over is assumed, which of the following would be the best statement regarding his hypothesis?

1) These results support his hypothesis.
2) The results would support his hypothesis if identical observations were made in a second cross.
3) These results neither support nor contradict his hypothesis.
4) These results contradict his hypothesis.
5) These results would contradict his hypothesis if identical results were obtained in a second cross.

19. In Question 18, if two of the short plants with yellow pods in the 2nd generation were crossed, what percentage of the offspring would be heterozygous for at least one trait?

1) 0%
2) 25%
3) 50%
4) 75%
5) 100%

20. A pet shop owner who specialized in breeding smooth coated white guinea pigs decided to introduce a new line of guinea pigs. He felt that a rough coated white animal would be very popular and to develop such an animal he ordered a pair of guinea pigs from another breeder who specialized in dark colored, rough coated animals. The pet shop owner then crossed the new dark rough male with one of his smooth, white females. All of the offspring from this cross had dark fur and rough coats. The shop owner was disappointed but set out to solve his problem by trying the five approaches listed below. Which one of the five would have the best chance of producing a rough coated, white animal?

1) Crossed the newly purchased dark rough female with one of his smooth white males.
2) Crossed a dark rough coated male offspring with a smooth white female.
3) Crossed a dark rough coated male offspring with a rough coated female offspring.
4) Crossed two smooth white guinea pigs.
5) Crossed the pair of dark rough animals he had obtained from the other breeder.

21. A geneticist, befriended by the pet shop owner, became interested in the guinea pig problem. The question of interest to the geneticist was whether or not the genes controlling coat color and coat texture were linked to the same pair of chromosomes. The geneticist suspected that the genes were indeed linked and he set out to test his hypothesis. He crossed the dark rough coated male purchased by the pet shop owner with a smooth white female. Of course all of the offspring were dark rough coated animals. The geneticist now mated two of these offspring and while waiting on the arrival of the young guinea pigs he made a prediction in keeping with his hypothesis. Which of the following would best represent that prediction?

1) All offspring will be dark rough.
2) One half of the offspring will be white smooth and one half will be dark rough.
3) 25% will be dark rough, 25% dark smooth, 25% white rough and 25% white smooth.
4) One fourth will be white smooth and 3/4 will be dark rough.
5) Nine sixteenths will be dark rough, 3/16 will be dark smooth, 3/16 will be white rough and 1/16 will be white smooth.

22. Suppose the geneticist chose the alternate hypothesis, that the genes for coat color and texture are located on two different sets of homologous chromosomes. Which of the following predictions would be consistent with this hypothesis?

1) All offspring will be dark rough.
2) One half of the offspring will be white smooth and one half will be dark rough.
3) 25% will be dark rough, 25% dark smooth, 25% white rough and 25% white smooth.
4) One fourth will be white smooth and 3/4 will be dark rough.
5) Nine sixteenths will be dark rough, 3/16 will be dark smooth, 3/16 will be white rough and 1/16 will be white smooth.

23. A couple consulted a genetic counselor before having children. By analysis of tissue samples and family histories the counselor determined that both partners carried defective genes for two critical enzymes X and Y. The genes for X and Y are located on separate non-homologous chromosomes. Individuals that are homozygous for either defective gene show abnormal development and die before reaching maturity. What is the number of different kinds of gametes that would be produced by either parent, that is, gametes that differ with respect to the 2 different genes for enzymes X and Y?

1) one
2) two
3) three
4) four
5) six

24. If the couple in the previous problem insisted upon having a child, what would be the probability of the couple producing a child showing abnormal development because of defective X or Y enzymes?

1) 1/4 (one chance in four)
2) 1/8
3) 1/2
4) 7/16
5) 9/16

25. A biologist was studying an organism with only two pairs of homologous chromosomes. This biologist was also studying the inheritance of the following three traits:

height — tall or short
color — blue or green
weight — heavy or light

Which of the following could you most confidently conclude about this organism?

1) The tall, blue, and heavy traits are dominant.
2) At least two of the traits are linked.
3) Even with crossing-over, these organisms could produce only two different kinds of gametes.
4) The tall, blue, and heavy traits are recessive.
5) All the above can be concluded with equal confidence.

26. A mouse is heterozygous dark-albino and homozygous for short ears. How many kinds of gametes can it produce?

1) Two.
2) Two if these genes are linked, four if they are not.
3) Four if these genes are linked, two if they are not.
4) Four.
5) Two if these genes are linked, four if they are not and four if crossing-over occurs.

27. The mouse in Question 26 (heterozygous dark-albino, homozygous short ears) is crossed with a mouse which is heterozygous dark-albino and homozygous for long ears. How many different genotypes can there be among the offspring?

1) Two
2) Three
3) Four
4) Eight
5) Two if these genes are linked, four if they are not.

28. How many phenotypes can there be among the offspring in question 27?

1) Two
2) Three
3) Four

34

4) Eight
5) Two if by land, four if by sea.

29. Suppose you cross a long-furred white mouse with a short-furred black mouse. All of the offspring are short-furred and black. You cross several of these offspring with long-furred white mice. You find that of 75 offspring, one of them is white with short fur, 37 are black with short fur, and 37 are white with long fur. Which of the following is the best explanation for these observations?

1) An egg or a sperm carried genes for short fur and white color.
2) Crossing-over probably occurred.
3) The genes for fur color and fur length were linked.
4) 1, 2 and 3 all contribute to an explanation.
5) 2 and 3 are the best explanations.

30. A student was considering how two traits (such as fruit color and leaf shape) might be assorted in the gametes of an organism. He said that it would not matter whether the genes were carried on one pair or on two pairs of homologues. In which of the following cases would this student's statement be valid?

1) The organism is heterozygous for both traits.
2) The organism is heterozygous for one trait but homozygous for the other.
3) The organism is homozygous for both traits.
4) Both 1 and 2.
5) Both 2 and 3.

31. A crab may have either 4 or 5 spines on its first walking legs and it may have either red or black bumps on its claws. Suppose you cross a red-bumped, 4-spined female with a black-bumped and 5-spined male. All of the offspring have red bumps and 5 spines. Which of the following genotypes best represents one of these offspring?

1) Homozygous for bump color and spine number.
2) Homozygous for bump color, heterozygous for spine number.
3) Heterozygous for bump color, homozygous for spine number.
4) Heterozygous for bump color and spine number.
5) Homozygous dominant, heterozygous recessive.

32. If you crossed the red-bumped, 5-spined offspring with a black-bumped, 5-spined individual, genetically identical to the male in the last question, which of the following would be the best statement about the offspring (the second generation)?

1) The frequency of phenotypes would be the same whether the genes for bump color and spine number are linked or not.
2) If these genes are not linked, four different phenotypes would be observed.
3) All the second generation offspring will have black bumps.
4) If these genes are linked, 50% of the second generation offspring will have five spines and 50% will have four spines.
5) Both 2 and 4 are valid statements.

33. To test the hypothesis that the genes for bump color and spine number are linked, you could cross two of the first generation offspring from Question 32. Which of the following observations would be the best evidence in favor of the hypothesis?

1) This cross produces no individuals with black bumps and four spines.
2) Of the offspring from this cross, 75% have red bumps.
3) Of the offspring from this cross, 25% have four spines.

4) Among the offspring from this cross, the commonest phenotype is red-bumped and five-spined.
5) There are only four phenotypes present among the offspring.

34. A fish farmer produced rainbow trout for stocking by private land owners and fishing clubs. Among the many normal dark colored rainbow trout he produced, there were always a few gold-colored trout which grew to a larger size than the normal trout. He wanted to get a pure strain for production and bred a gold female to a normal dark male. The offspring were all dark-colored. What would be the most likely result of a cross between a dark male offspring and a gold female?

1) 50% would be dark and 50% would be gold.
2) 75% would be gold and 25% would be dark.
3) 75% would be dark and 25% would be gold.
4) They would all be dark.
5) They would all be gold.

35. With reference to the trout in the preceding question, when the first cross was attempted between the gold female and the normal dark male, an unexpected result was obtained because the X-chromosomes of gold trout were incompatible with X-chromosomes of normal dark trout. Fertilized eggs with a combination of incompatible X-chromosomes failed to develop. All combinations of X and Y chromosomes developed normally. What would be the most likely sex ratio of the attempted cross?

1) No males: all females
2) 2 males: 1 female
3) 1 male: 1 female
4) 1 male: 2 females
5) All males: no females

36. The fish farmer was eventually successful in establishing a strain of gold rainbow trout which grew rapidly when fed a dry food supplemented with a certain vitamin. When he changed to a food which did not contain the vitamin, the strain of gold trout did not survive. The dark trout did not require the vitamin and grew well on the new diet.

He crossed a dark trout with a gold trout. All of the offspring grew without the vitamin. He made the hypothesis that the genes for color and those relating to the vitamin were linked on one set of homologous chromosomes. If he bred a dark offspring with a gold trout, which of the following predictions would he most likely make.

The vitamin will be required by:

1) 25% of the gold trout and 75% of the dark trout.
2) 50% of the gold trout and 50% of the dark trout.
3) 75% of the gold trout and 25% of the dark trout.
4) 100% of the gold trout and none of the dark trout.
5) All of the trout, gold and dark.

37. If the trout farmer had hypothesized that the genes for color and those relating to the vitamin were on two different sets of homologous chromosomes, which of the following would he most likely predict about the offspring in the previous question?

The vitamin would be required by:

1) 25% of the gold trout and 75% of the dark trout.
2) 50% of the gold trout and 50% of the dark trout.
3) 75% of the gold trout and 25% of the dark trout.
4) 100% of the gold trout and none of the dark trout.
5) All of the trout, gold and dark.

38. When crossing-over occurs, chromosomes are changed in specific ways, creating combinations which would not be observed before crossing over. Which of the following would you most likely find only as a result of crossing-over?

   1) Two homologous chromosomes have information for different traits.
   2) Two homologous chromosomes have identical information.
   3) One chromosome contains two different expressions of the same trait.
   4) One chromosome has information for one expression of a trait.
   5) Both 2 and 4 would be equally likely.

39. Hemophilia is the "bleeder's" disease — people who have it have blood which does not coagulate well and they die early. The gene for the disease is recessive and sex-linked. Which of the following would be the best prediction about the offspring of a hemophiliac male and a non-hemophiliac female?

   1) All children would be hemophiliac.
   2) All females would be hemophiliac.
   3) All males would be hemophiliac.
   4) Hemophiliac children would be in a 3:1 ratio, normal to hemophiliac.
   5) None of the above is a good answer.

40. For this question only assume that the mother in the previous question was a carrier, i.e., she had the hemophilia gene on one of her chromosomes, but the other was normal. Which of the following would now be the best prediction about the offspring?

   1) All males would be hemophiliac.
   2) All females would be hemophiliac.
   3) Half of all offspring (whether male or female) would be hemophiliac.
   4) Half the males would be hemophiliac but no females would be.
   5) Half the females would be hemophiliac but no males would be.

41. In one family there were three brothers and all were color blind. Suppose the father and mother had a daughter as a 4th child. Which of the following inferences could you most confidently make about the color vision of this child?

   1) Even if the father is colorblind, she will have normal vision.
   2) If the father has normal vision, she will have normal vision.
   3) She will be colorblind.
   4) If the mother is colorblind, the daughter will be colorblind.
   5) Either 2 or 4 are equally likely choices.

42. In *Drosophila* notched wings are a dominant sex-linked trait. If a normal wing male is mated with a heterozygous notched female what percentage of the female offspring will have notched wings?

   1) 0%
   2) 25%
   3) 50%
   4) 75%
   5) 100%

43. From Question 42, what percentage of the male offspring will have notched wings?

   1) 0%
   2) 25%
   3) 50%

4) 75%
5) 100%

44. What percentage of the gametes produced by the original male in Question 42 will carry the gene for the normal expression?

1) 0%
2) 33 1/3%
3) 66 2/3%
4) 50%
5) 100%

45. In birds, in contrast to mammals, the females have one X and one Y sex chromosome and the males have two X chromosomes. With canaries, the green variety with black eyes is dependent on a product of a sex-linked gene. Canaries that lack this product are the cinnamon variety with red eyes. What is the expected appearance of the progeny from a cross between a cinnamon male and a green female?

1) 100% of the males will be cinnamon and 100% of the females green.
2) 100% of the males will be green and 100% of the females cinnamon.
3) 100% of the males will be green and 50% of the females cinnamon.
4) 50% of the males will be green and 100% of the females cinnamon.
5) All of the offspring will be green.

46. In Plymouth Rock chickens barred plumage is dominant over non-barred and this trait is located on the sex chromosomes. If a homozygous barred male is mated to a non-barred female, if parents do not mate with offspring, and if the original male and female are not counted as a generation, how many generations will it be until a non-barred male is produced?

1) One
2) Two
3) Three
4) Four
5) Five

47. A family pussycat is permitted to leave the house each night at its will. Over the years it produces 79 kittens. The owner notes that the cat has had 26 male kittens and 53 female kittens. Cats, on the average, produce equal numbers of both sexes. What explanation of this apparent discrepancy is most likely?

1) The cat always mates with the same male that happens to produce only female gametes.
2) The cat carries a recessive lethal (kills zygote or young embryo) trait that is not sex-linked.
3) The cat carries a dominant lethal trait that is not sex-linked.
4) The cat carries a recessive lethal trait that is sex-linked.
5) The cat carries a dominant lethal trait that is sex-linked.

48. The three sexes of the Dudleys are represented by three assortments of sex chromosomes; Flubs have three X chromosomes; dubs have two X's and a Y; and suds have an X and two Y's. One of the genes on the X chromosome codes for a vanilla flavor when only one chromosome carries the gene, and for a maple flavor when two chromosomes each carry an active gene. Three active genes make the Dudley carrying them taste like burnt toast. If a maple-flavored flub mated with a vanilla dub and a vanilla sud, how many maple-flavored suds would be included among the offspring?

1) 8
2) 6
3) 4
4) 2
5) 0

49.   A blood analysis of a small child indicated two different forms (R and S) of a particular sugar-digesting enzyme were present. The two forms differed in only a few amino acid positions and both forms were active. Forms R and S were present in approximately equal amounts in the blood sample used for analysis. Assuming that information for this enzyme is carried on a single homologous pair of chromosomes, which of the following statements is most justified?

1)   At least one parent possessed genetic information for form S.
2)   Neither parent can be homozygous for form R or form S.
3)   It is possible that neither parent possessed genetic information for form R.
4)   At least one parent is homozygous for form S.
5)   Choices 2 and 3 are equally likely as the best choice.

50.   In Question 49, suppose it was later discovered that the child's father produced only form S of the enzyme. Which of the following would be the best statement regarding the genetic information for the enzyme possessed by the child's mother?

1)   She would also be homozygous for form S.
2)   She would be homozygous for form R.
3)   It is possible that the mother has no genetic information for either form of the enzyme.
4)   She would be heterozygous, i.e., have genetic information for both forms of the enzyme.
5)   Choices 2 and 4 are equally likely as the best statement.

51.   Suppose the parents described in Questions 49 and 50 had another child. Which of the following would be the most likely prediction regarding the genetic information for the sugar digesting enzyme possessed by this second child?

1)   The child would be homozygous for form R.
2)   The child would be homozygous for form S.
3)   If crossing-over occurs the child would have no information for either form.
4)   The child would be heterozygous, i.e. have genetic information for both forms.
5)   Choices 2 and 4 are equally likely as the best statement.

52.   Imagine that the doughnut is a sexually reproducing animal and that there is a single enzyme that forms the hole. All doughnuts observed in the past have holes. Suppose a mutation occurs in one gene of one doughnut so that it now codes for an inactive protein instead of the hole-forming enzyme. Which of the following predictions would be most likely in descendants of this mutant doughnut?

1)   The next generation will all have holes.
2)   Heterozygous doughnuts would not have holes.
3)   If the cell in which the mutation occurs undergoes mitosis, subsequent generations of doughnuts will all have holes.
4)   The next generation will all be holeless.
5)   Both 1 and 3 are valid predictions.

53.   Suppose that the doughnut in which the mutation for holelessness occurs is heterozygous for glaze and powdered sugar coating. If the genes for surface coat are on the same chromosome as the gene for hole-forming enzyme, which of the following is most likely?

1)   Unless crossing-over occurs, there will be no glazed doughnuts without holes.

   2)   If powdered sugar is dominant, it will be linked to holelessness.

   3)   If crossing-over occurs, there will be no doughnuts heterozygous for both surface coat and holes.

   4)   Unless crossing-over occurs, there will be no holeless doughnuts which are also heterozygous for surface coat.

   5)   Holeless doughnuts will all be glazed.

54.   A biologist was studying a plant (the helix widget) which produced spores. Each of these spores grew into a multicellular body (A) which produced gametes. The zygotes resulting from fusion of gametes from different plants grew into a second multicellular body (B) which produced spores. Information for color in this plant was known to be carried on a single homologous pair of chromosomes and could be expressed as either green or brown. Brown was dominant. She collected spores from a plant known to be heterozygous for this color trait and allowed the spores to grow into the multicellular body (A). What percentage of these multicellular bodies will be brown?

   1)   100%
   2)   75%
   3)   50%
   4)   25%
   5)   0%

55.   Suppose the biologist in Question 54 now combined gametes from brown multicellular bodies (A) with gametes from green multicellular bodies (A). What percentage of the resulting multicellular bodies (B) would be green?

   1)   100%
   2)   75%
   3)   50%
   4)   25%
   5)   0%

GENE EXPRESSION AND PROTEIN SYNTHESIS

Deoxyribonucleic acid (DNA), a major component of chromosomes, carries genetic information for the traits of living cells and organisms. DNA is a polymer composed of four nucleotide monomers, each containing one of the nitrogenous bases adenine (A), guanine (G), thymine (T), or cytosine (C). The DNA molecule is self-replicating, a necessary property of the material carrying and transmitting genetic information. Replication is possible because each nitrogenous base pairs specifically (base pairing) with one other base, A with T and C with G. Thus one strand can specify a complementary sequence of monomers in a second DNA strand or polymer.

The sequence of monomers in DNA also directs protein synthesis with the participation of ribonucleic acid (RNA). The RNA polymer is composed of four nucleotide monomers, three of which are also found in DNA but the fourth, uracil (U), replaces T. In the initial transcription event of gene expression messenger RNA (mRNA) is formed by base pairing with a small portion of the DNA in the nucleus, thus the sequence of monomers in the mRNA is complementary to the sequence on the DNA. Another type of RNA, transfer RNA or tRNA, is also transcribed from the DNA. Both types of RNA move from the nucleus into the cytoplasm. Each set of three monomers in the mRNA is a codon and pairs with a complementary set of bases (anti-codon) in a tRNA molecule. Before pairing with the mRNA, each tRNA has formed a chemical bond with a specific amino acid. The result of codon-anti-codon pairing is the arrangement of a specific sequence of amino acids in a row. The bonds between tRNAs and amino acids are broken and new bonds are formed between adjacent amino acids. Thus in translation the sequence of anti-codons uniquely specifies a sequence of amino acids, which become chemically bonded to form a protein molecule.

Very little of the information in the DNA is expressed at any one time, indicating that some regulating mechanism must exist. Masking of unused DNA may be accomplished by protein in combination with the DNA. Some observations have disclosed localized "puffing" in giant insect chromosomes, apparently when mRNA is being synthesized. Such chromosomal puffing seems to be under the control of other substances such as hormones. Similarly, both internal and environmental factors seem to influence the regulated expression of genes in all organisms.

EXAMPLE 1

Human hemoglobin differs significantly in amino acid sequence from frog hemoglobin. When human hemoglobin messenger RNA (mRNA) is carefully injected into frog eggs traces of human hemoglobin can be found in the eggs after one hour. From this experiment it can be concluded that:

1) Uninjected frog eggs must contain the genes for human hemoglobin.
2) The same code words (groups of 3 nucleotides on mRNA) specify the same amino acids during translation in humans as in frogs.
3) The rate of RNA synthesis (transcription) must be very high in uninjected frog eggs.
4) Human hemoglobin genes are dominant over frog hemoglobin genes.
5) All of the above conclusions are valid.

ANALYSIS:

1. Genetic information in DNA is not required in this case since mRNA has been injected and thus the genes are no longer needed to direct protein synthesis. — Reject

2. Since no RNA other than mRNA was injected, frog tRNA must be used in hemoglobin synthesis. The frog tRNA must (with appropriate amino acids attached) be matching with code words on the human mRNA. If the code words in humans and frogs differed, then the amino acid sequence would differ and human hemoglobin would not be produced. — Accept.

3. Irrelevant — the issue here is the synthesis of a protein with the correct sequence of amino acids — not how rapidly RNA is synthesized. Whether the rate of RNA synthesis were faster or slower in uninjected eggs would make no difference in this problem.

4. Although one might intuitively think that human hemoglobin genes are dominant since they are expressed, the problem does not state that frog hemoglobin is not synthesized. More importantly, a more meaningful conclusion about the coding of information and protein synthesis in frogs and humans can be drawn from the information provided in this problem (see number 2).

5. Reject - choices 1, 3 and 4 were rejected.

EXAMPLE 2

A mouse is heterozygous for light and dark fur. It has dark fur. Which of the following statements is the best explanation of this observation?

1) The gene for dark fur codes for an enzyme which destroys the gene for light fur.
2) The gene for dark fur is dominant.
3) The mouse has one copy of the genetic information for a protein which does not catalyze the formation of melanin.
4) The mouse has one copy of the genetic information for a protein which catalyzes the formation of melanin.
5) The mouse has one copy of the genetic information for a protein which catalyzes the formation of white pigment in fur.

ANALYSIS:

1. This mechanism could account for the absence of white fur in the heterozygous individual. If the gene for white fur is destroyed, then it obviously will not be expressed. However, this

mechanism does not account for the production of brown color unless it is assumed that the enzyme also catalyzes that reaction - an unlikely condition. Furthermore, destruction of the gene is contradicted by the observation that heterozygous mice can produce white offspring.

2.   Although it is immediately obvious that dark fur is the dominant trait, the problem asks for the best explanation of the observations. Merely noting that dark is dominant describes the system without providing an explanation.

3.   This statement is probably true but it does not explain the presence of melanin in the fur. By itself, this statement would lead to the prediction that the mouse would be white.

4.   If the mouse has the information and we assume it is expressed, then the enzyme will be present, melanin will be formed and the mouse will have dark fur. This is the best answer.

5.   If this statement was true, the mouse would have white pigment in its fur and would appear white or intermediate in color between a white mouse and one which had only melanin. This contradicts the observation.

1. A biochemist was studying a sexually reproducing organism that normally produced a red pigment. He found a few of the organisms without the red pigment, and suspected that there had been a change in the DNA in the individuals. Which of the following procedures would provide the best evidence as to whether the organisms have altered DNA?

    1)   Search for more specimens without pigment.
    2)   Identify the enzymes that lead to the production of the pigment.
    3)   Identify the compounds that make up the pigment and determine if they are all present in the pigment-free organisms.
    4)   Breed pigment-producing (normal) organisms with each other to see if a pigment-free organism arises by mutation after several generations.
    5)   Breed pigment-producing (normal) organisms with non-pigment producing organisms and examine the pigment producing abilities of the progeny after several generations.

2. The nucleus is removed from an egg cell laid by frog A. Into the enucleated egg cell is transplanted a nucleus from an intestine cell of frog B. The cells of frog A contain protein A, but the cells of frog B do not. The egg with the transplanted nucleus grows and divides several times, then the cluster of cells is analyzed. Protein A is found in the cells. Which of the following is the best interpretation?

    1)   The transplanted nucleus contains information for protein A.
    2)   Some mRNA, synthesized on the gene for protein A, was left in the cytoplasm when the egg nucleus was removed.
    3)   Some tRNA, synthesized on the egg cell DNA, was left in the cytoplasm when the egg nucleus was removed.
    4)   The synthesis of protein A is directed by information contained in enzymes in the egg cytoplasm and not in DNA or RNA.
    5)   Both 2 and 3 are equally valid.

3. Now suppose it is discovered that the egg cell had been fertilized by sperm from frog C before its nucleus was removed. Cells of frog C contain protein A. Which of the following is the best interpretation?

    1)   The transplanted nucleus contains information for protein A.
    2)   Some mRNA, synthesized on the gene for protein A, was left in the cytoplasm when the egg nucleus was removed.
    3)   Some tRNA, synthesized on the egg cell DNA was left in the cytoplasm when the egg nucleus was removed.
    4)   The synthesis of protein A is directed by information contained in the enzymes in the egg cytoplasm and not in DNA or RNA.
    5)   Both 2 and 3 are equally valid.

4. Another egg from frog A is fertilized by sperm from frog C. A little later it is enucleated and receives a transplant of an intestine cell nucleus from frog B. Just before transplantation, however, a chemical is added which prevents the synthesis of RNA. After several cell divisions the cluster of cells contains protein A. Which of the following is the best interpretation?

    1)   The transplanted nucleus contains information for protein A.
    2)   Some mRNA, synthesized on the gene for protein A, was left in the cytoplasm when the egg nucleus was removed.
    3)   Some tRNA, synthesized on the egg cell DNA, was left in the cytoplasm when the egg nucleus was removed.
    4)   The synthesis of protein A is directed by information contained in enzymes in the egg cytoplasm and not in DNA or RNA.
    5)   Both 2 and 3 are equally valid.

5.   Consider the hypothesis that RNA is synthesized on DNA. To test this hypothesis some investigators injected radioactive monomers into organisms A and B so that radioactive RNA was formed. They purified the labelled RNA from each organism. They also purified the DNA from organism A, heated it to cause the strands to separate and chemically attached the single strands to plastic beads. When these beads were then mixed with RNA from organism A, radioactivity stuck to them. When they were mixed with radioactive RNA from organism B, it did not stick to them. Many scientists accept observations such as these as evidence that RNA is synthesized on DNA. Which of the following assumptions are they making?

   1)   Radioactivity causes mutations in organism B.
   2)   Heating DNA changes the sequence of its monomers.
   3)   RNA molecules from organisms A and B have the same sequence of monomers.
   4)   Only nucleic acids with complementary sequences of monomers can stick together by base pairing.
   5)   Only nucleic acids with the same sequence of monomers can stick together by base pairing.

6.   A scientist wishes to study the synthesis and movement of nucleic acids in some insect cells. The cells are placed into a growth medium containing radioactive nucleotide "U" (uridine). For the duration of the experiment the concentration of radioactive uridine in the medium is kept constant. At 10 min. intervals, small samples of cells are removed and the unincorporated uridine is washed away with acid. The radioactivity incorporated into nucleic acid is located. If these cells behave like other eucaryotic cells, one would expect to find radioactivity appearing:

   1)   First in the cytoplasm and later only in the nucleus.
   2)   First in the nucleus and later only in the cytoplasm.
   3)   First in the cytoplasm and later in both cytoplasm and nucleus.
   4)   First in the nucleus and later in both cytoplasm and nucleus.
   5)   In the nucleus only at all times sampled.

7.   The scientist repeats the experiment described in the previous problem using radioactive monomer "T" (thymidine) in place of uridine in the culture medium. If these insect cells behave like other eucaryotic cells, one would now expect radioactivity appearing:

   1)   First in the cytoplasm and later only in the nucleus.
   2)   First in the nucleus and later only in the cytoplasm.
   3)   First in the cytoplasm and later in both cytoplasm and nucleus.
   4)   First in the nucleus and later in both cytoplasm and nucleus.
   5)   In the nucleus only at all times sampled.

8.   Before human red blood cells are released into the bloodstream they lose their nuclei. Yet it can be shown that for a considerable time after they are in circulation they are capable of synthesizing additional hemoglobin. That is, if red blood cells without nuclei are incubated in a medium containing radioactive amino acids, some of the labeled amino acids are incorporated into hemoglobin. Based upon what you know of transcription and translation of genetic information, which of the following is most justified?

   1)   Hemoglobin must be capable of self replication.
   2)   A portion of the nucleus containing the hemoglobin genes remains behind when the rest of the nucleus is lost from the developing red blood cells.
   3)   Some of the RNA that encodes the information for hemoglobin remains in the cytoplasm after the nucleus is lost.
   4)   Hemoglobin is not translated directly from nucleic acid but is made by enzymes that were synthesized before the cell nuclei are lost.
   5)   Answers 2 and 3 are equally justified.

9. When actinomycin, a known inhibitor of RNA synthesis, is added to the medium containing the amino acids and red blood cells described in problem 8, there is no change in the rate of hemoglobin synthesis. This additional information:

    1) Supports answer #1 in problem 8.
    2) Supports answer #2 in problem 8.
    3) Supports answer #3 in problem 8.
    4) Supports answer #4 in problem 8.
    5) Supports answer #5 in problem 8.

10. The color (pigment) in the feathers of certain birds is governed by a single pair of genes, one for dark and one for light feathers. It is also known that any given individual may appear dark at one time of year and light at another time of year. Which of the following hypotheses can be drawn from these observations?

    1) The environment influences the expression of genetic information.
    2) Genetic information is expressed sequentially and can be dormant (not expressed) for a period of time after which it can be reactivated and expressed again.
    3) The enzymes that catalyze formation of pigment are only produced at certain times of the year.
    4) The precursor substances necessary for formation of the pigment are available only at certain times of the year.
    5) All of the above are equally valid.

11. In the Arboretum two students observed bluebells growing in two different locations (A and B). At A all the plants had pink flowers and at B all plants had blue flowers. Student X hypothesized that the soil types at A and B were different and this could account for the differences in flower color. Student Y hypothesized that plants at A were genetically different from those at B. They decided to do an experiment. They produced clones of both plants. Plants from clone A were planted at locations A and B and plants from clone B were planted at locations A and B. Their experiment would:

    1) Be unnecessary since air temperature is known to affect flower color in bluebells.
    2) Test the hypothesis of student X only.
    3) Test the hypothesis of student Y only.
    4) Test the hypotheses of both students.
    5) Not test their hypotheses.

12. While studying butterflies in some South American mountains a biologist observes one kind of butterfly with large solid blue wings at high elevations. At lower elevations he observes what appears to be the same kind of butterfly but with much smaller wings and a thin black border on the wing edges. He states that the butterflies at high and low elevations are genetically the same and that wing differences are the result of environmental differences. Which of the following is the best description of his statement?

    1) It is a hypothesis which cannot be tested.
    2) It is a conclusion supported by his observation.
    3) It is a hypothesis which can be tested.
    4) It is a conclusion contradicted by his observations.
    5) His statement could be either 1 or 2 depending on how he analyzes the data.

13. The biologist in Question 12 later makes exactly the same observations on butterflies in a mountain range further to the east. These new observations:

    1) Strengthen the validity of his previous statement.
    2) Will force him to change his previous statement.

3) Can be regarded as data supporting his previous statement.
4) Weaken the validity of his previous statement.
5) Have no effect on the validity of his previous statement.

14. Later, the biologist in Questiton 12 was visiting still a third range of mountains. Here he observed the butterflies with large solid blue wings and the butterflies with smaller blue wings edged in black occupying the same areas in a high valley. These observations:

1) Strengthen the validity of his previous statement.
2) Force him to regard his previous statement as a conclusion.
3) Can be regarded as data supporting his previous statement.
4) Weaken the validity of his previous statement.
5) Have no effect on the validity of his previous statement.

15. Suppose you know:

A. A frog intestine cell nucleus can be transplanted into a frog egg which has had the egg nucleus removed. The egg can then grow into a normal mature frog.
B. Observations of different cells, such as nerve and muscle, indicate that they are very different in structure and function.

These two observations together suggest that:

1) Cells can change structurally if part or all of a chromosome is lost.
2) Mutations arising in cells can account for these observations.
3) The accurate copying of DNA is critical to insure precise transfer of genetic information.
4) Any cell is using only part of its genetic information.
5) The environment of an intestinal cell is different from that of an egg.

16. A biologist transplanted a nucleus from a frog intestine cell into an egg cell of another frog after the nucleus was removed from the egg cell. He did this with 100 pairs of cells. After incubation 72 eggs died, 26 eggs developed into abnormal frogs and 2 eggs developed into normal, complete frogs. He concluded that frog intestine cell nuclei contain all the information necessary to direct development of a whole frog. Which of the following assumptions did he make about the 98 failures?

1) They were damaged by the transplanting process.
2) Their nuclei did not contain all the information for a whole frog.
3) They contained the haploid number of chromosomes.
4) Their nuclei contained three haploid sets of chromosomes.
5) If they had been kept cooler, they would have developed.

17. If the intestine cells from which the nuclei were transplanted in Question 22 had contained an enzyme not found in intestine cells of the frog which donated the eggs, what would you predict about the intestine cells of the two frogs which developed from the incubated eggs?

1) They would not contain the enzyme.
2) They would contain the enzyme, but only half as much.
3) They would contain the enzyme.
4) They would contain twice as much of the enzyme.
5) 1 and 3 are equally likely.

18. A geneticist hypothesized that chromosome "puffs" represented parts of the chromosome where information was being used for protein synthesis by the cell. This geneticist had devised a way to directly observe newly synthesized mRNA. Which of the following observations on mRNA would best support his hypothesis?

1) mRNA present throughout the cell.
2) mRNA being synthesized at the chromosome "puffs" and at other parts of the chromosome.
3) mRNA being synthesized only at the chromosome "puffs".
4) mRNA being synthesized only on parts of the chromosome where "puffs" do not occur.
5) Choices 1 and 2 support his hypothesis equally well.

19. Suppose you discovered that an average-sized protein was composed of 400 monomers, from your knowledge of DNA involvement in protein synthesis, how big would you expect an average gene to be?

1) 400 nucleotides (monomers)
2) 800 nucleotides (monomers)
3) 1200 nucleotides (monomers)
4) 1600 nucleotides (monomers)
5) 2, 3, or 4 nucleotides (monomers)

20. Assuming DNA is transcribed to mRNA and this is translated into protein, which of the following sequences of DNA would match an amino acid sequence of Pro-Leu-Glu-His?

Amino Acid	mRNA code
His	CAC
Leu	CUC
Pro	CCA
Glu	CAA

1) CCA-CUC-CAA-CAC
2) GGT-GAG-GTT-GTG
3) GGU-GTG-GUU-GUU
4) CAC-CAA-CUC-CCA
5) CCU-CTC-CUU-CUC

21. A portion of a protein polymer from an organism contained the amino acid sequence: -Thr-Thr-Lys-. Which of the following sequences of DNA monomers most likely contains the information specifying this sequence of amino acids?

1) ATC-ATC-GCA
2) CTA-ATC-GCA
3) CTA-ATC-CAT
4) TCA-CAT-TCA
5) Both 2 and 3 are equally likely.

22. A plant physiologist was examining the kinds of protein produced in two closely related plants. He found that some of the proteins were different in the two plants. Which of the following must also be different in the two plants?

1) DNA
2) amino acids
3) mRNA
4) transfer RNA
5) Both 1 and 3 must be different in the two plants.

23. A scientist has isolated some RNA from a cell and wishes to determine the base sequence of the DNA from which the RNA was synthesized. Which of the following must be known?

1) Base sequence in RNA

2) Location of protein synthesis.
3) Which RNA bases pair with which DNA bases.
4) Both 1 and 2.
5) Both 1 and 3.

24. If nucleotide C coded for four amino acids identified as Val, Gly, Leu and Ser and the other three nucleotides each coded for four different amino acids, which of the following would represent a serious problem in protein synthesis?

1) There would not be enough nucleotides to code for the twenty amino acids occurring in an organism.
2) There would be no way for messenger RNA to be transcribed from the DNA.
3) There would be more than one sequence of amino acids for any given sequence of nucleotides.
4) There cannot be a three to one correspondence between nucleotides and amino acids.
5) Both 1 and 3 would be problems.

25. If each nucleotide in DNA paired only with itself (A and A, C with C, etc.), which of the following statements would be most likely?

1) A pairs with T and C pairs with G.
2) DNA could not replicate.
3) The two strands of each DNA molecule would be identical.
4) DNA could not carry genetic information.
5) More monomers would be required to synthesize DNA.

26. A student states, "If A paired with G and C paired with T, DNA could not be replicated." How would you criticize this statement?

1) A pairs with T and C pairs with G.
2) As long as each nucleotide pairs specifically with only one other, replications will occur.
3) The two strands must separate before replication can occur.
4) If A paired with C and G paired with T, DNA could not be replicated.
5) He is assuming that DNA carries the information specifying a cell's characteristics.

27. A scientist performs an analysis on two samples of protein to determine their amino acid content and obtains the following results:

Amino Acid	Sample A	Sample B
glycine	10%	10%
methionine	5%	5%
serine	40%	40%
proline	35%	35%

He concludes that the two samples are composed of the same protein. How would you criticize his conclusion?

1) The samples may have come from different organisms.
2) There are too few amino acids composing the protein in the two samples.
3) In both samples the protein was composed of the same type of amino acids.
4) The amino acids from the two protein samples may have had different sequences.
5) You have no way of criticizing his conclusion.

28. Suppose you discover an organism containing only six amino acids. Which of the following inferences about this organism would be most justified?

1) This organism produces only six different forms of messenger RNA.
2) Amino acids must be coded by sets of at least two nucleotides in DNA in this organism.
3) The organisms cannot produce more than six different kinds of protein.
4) The protein of this organism will contain monomers other than amino acids.
5) This organism could produce protein molecules no more than six amino acids in length.

29. Suppose you discover an organism in which all proteins are composed of only 13 amino acids. Which one of the following predictions would be least acceptable?

1) Amino acids will be coded by sets of two nucleotides in DNA.
2) There will be 13 different tRNA's.
3) mRNA will be transcribed from DNA.
4) Amino acids will be coded by single nucleotides in DNA.
5) There will be as many mRNA's as there are kinds of protein.

30. Suppose one strand of mRNA coded for a protein molecule 347 amino acid monomers in length. If the first 2 monomers were lost from the mRNA, how many of the amino acid monomers would be changed in protein molecules later synthesized using this information?

1) 1
2) 2
3) 4
4) 0
5) More than 4.

31. Suppose a biochemist made a strand of mRNA that was 300 monomers long. He also added an enzyme that randomly destroyed any 3 of the first 12 monomers. How many of the amino acid monomers would be expected to be changed in the resulting protein molecule that will be later synthesized?

1) All of the resulting 100 amino acids will be affected.
2) There will only be 97 amino acids polymerized.
3) Since the enzyme works at random it is not possible to know how many amino acids will be affected.
4) Any of the first 4 amino acids might be affected.
5) Each third amino acid will be affected.

32. Suppose you know the following:
   A. A certain kind of yeast produces protein H.
   B. One culture in every 10 produces an unusual form of protein H in which two specific adjacent amino acids are in reverse order.
   C. All of the protein H produced by the one in ten cultures is of the unusual type.

Which of the following explanations would be the most probable?
1) There was a coding error in the DNA.
2) There was an error in transcribing messenger RNA from transfer RNA.
3) Transfer RNA could not attach properly to the appropriate amino acids.
4) Messenger RNA did not reach the site of protein synthesis.
5) Transfer RNA was transcribed incorrectly from DNA.

33. A chemical was added to a cell and it was observed that the rate of protein synthesis decreased. From your knowledge of protein synthesis which of the following is the best inference that can be made?

1) The chemical is an enzyme that catalyzes messenger RNA production.

2) The chemical destroys the nuclear membrane.
3) The chemical prevents the production of messenger RNA.
4) The chemical prevents the production of transfer RNA.
5) Both 3 and 4 are equally good.

34. A bacteriologist was studying the production of a purple pigment by some bacteria. He knew the pigment was synthesized in the following way:

Chemical A → chemical B → purple pigment

The bacteriologist isolated a mutant which would not produce the pigment. He then found that these mutant bacteria could produce the pigment if they were grown on a culture medium which contained chemical B. Which of the following could you most likely conclude about these mutant bacteria?

1) They could not produce the enzyme which catalyzes the conversion of chemical B to pigment.
2) The enzyme which catalyzes conversion of chemical B to pigment is probably not a protein.
3) They cannot produce the enzyme which catalyzes the destruction of the purple pigment.
4) They could not produce the enzyme which catalyzes the conversion of chemical A to chemical B.
5) Choices 1 and 4 are equally likely.

35. During protein synthesis, tRNA molecules "guide" amino acids to the mRNA and position them in a specific sequence. Suppose a molecule of the amino acid cysteine, bonded to a tRNA molecule, was changed to the amino acid alanine. Where would you predict the tRNA bearing the altered amino acid would interact with the mRNA?

1) At the specific alanine site.
2) At the specific cysteine site.
3) At a site midway between the cysteine and alanine sites.
4) The "defective" tRNA would be rejected by the ribosome.
5) Such accidents cannot occur.

36. A scientist prepared an extract from bacteria of type A. He determined that it was pure DNA. He mixed it with a culture of bacteria of type B. After a while living bacteria in the culture had characteristics of type A. He concluded that DNA carried information for characteristics of type A bacteria. Which of the following would be most important to know in order to evaluate his conclusion?

1) How did he determine that his extract was pure DNA?
2) What are the characteristics of type A and type B bacteria?
3) How many type A bacteria appeared in the treated culture?
4) How many times did the experiment fail before it succeeded?
5) Do type A bacteria produce enzymes that break down DNA?

37. Suppose you had a supply of nucleotides that were made with special atoms so that each nucleotide weighed twice as much as a naturally occurring nucleotide. Suppose you allowed a cell to replicate its DNA once under conditions where the only nucleotides available to it were the special heavy ones. Assume the commonly accepted hypotheses for the structure and replication of DNA are true. What do you predict about the weight of the double-stranded DNA molecules in the treated cell?

1) All will have the normal weight.
2) All will weigh twice normal.
3) Half will have normal weight and half will weigh twice normal.

4) All will weigh 1.5 times normal.
5) The weight cannot be predicted because we do not know how many heavy nucleotides will be incorporated.

38. Dark mice possess the enzyme which can synthesize the pigment melanin. A student predicted that in mice homozygous for dark color there would be twice as much melanin-synthesizing enzyme present as in heterozygous mice. Which of the following assumptions is he most likely making?

1) mRNA contains the information for the melanin synthesizing enzyme.
2) Information on both homologues is expressed.
3) The mice must be homozygous to appear dark.
4) DNA carries the genetic information in mice.
5) The mice can be heterozygous and appear dark.

39. In a certain hypothetical bird, feather color requires an enzyme which is four amino acids long. There are three different forms, as follows:

Sequence	Pigment
Ala-Leu-Lys-Cys	Brown
Ala-Leu-Arg-Cys	Buff
Ala-Arg-Lys-Cys	Speckled

Further, the amino acids have the following codes in the RNA:

Amino Acid	RNA Code
Ala	GCA
Leu	UUA
Lys	AAA
Cys	UGU
Arg	AGA

If you examined the DNA sequence in a homozygous brown individual, which of the following sequences would you expect to find?

1) GCA-UUA-AAA-UGU
2) CGT-AAT-TTT-TCT
3) CGT-AAT-TTT-ACA
4) GCA-TTA-AAA-UGU
5) CGU-AAU-UUU-ACA

40. A scientist crossed a female bird who had only information for long wings with a male bird who had only information for short wings. All the offspring had long wings. A molecular biologist found that the protein products of these genes were as follows:

Allele	Protein
long	Leu-Tyr-Leu
short	Tyr-Leu-Tyr

The biologist also knew the mRNA codes for the following amino acids:

Amino Acid	mRNA
Leucine	CCU
Tyrosine	GAC

Which of the following DNA sequences would the biologist find in the offspring from the scientist's cross?

1) CCT-GAC-CCT
2) GGA-CTG-GGA
3) CTG-GGA-CTG
4) GAC-CCT-GAC
5) All the above will be found.

POPULATION GENETICS

The gene pool of a population is the total of all genes for all individuals in the population. For the different expressions of each specific trait in the population a gene frequency can be calculated. For example, if B and b are genes of two expressions of a trait and only these genes for that trait are present in the population, the gene frequency for B is B/ B +b. Gene frequency will remain constant over time if there is random breeding within the population, no mutation, no selection for one gene over the other, and no migration. If these conditions do not hold, and frequently they do not, the gene frequencies will most likely change.

Selection is constantly acting on a population, usually on combinations of traits, and new combinations are constantly being produced by sexual reproduction. Occasionally some single selecting factor, such as the application of DDT or an antibiotic, can result in rapid selection for a single expression such as resistance. When gene frequencies in several or many traits change drastically and/ or new genes arise through mutation, new species may arise which cannot interbreed with the parent population. This process is thought to occur by isolation of a part of the original population and the subsequent differential actions of mutation and selection on the isolated populations. Geographical barriers can initially isolate populations which can then change in different ways. Even if populations later intermix after some genetic change has occurred, interbreeding may be prevented by temporal, behavioral, or morphological barriers so that the species is maintained as a separate entity.

EXAMPLE 1.

The helix widget is a plant which reproduces sexually. One phase of its life cycle is a multicellular structure developed from a zygote. A biologist studied twenty individuals in this phase of their life cycle. Ten were green and ten were brown and he knew that brown is dominant. Which of the following is most likely to describe the gene frequencies for color in this population?

    1) 0.9 green, 0.1 brown
    2) 0.8 green, 0.2 brown
    3) 0.6 green, 0.4 brown
    4) 0.4 green, 0.6 brown
    5) 0.1 green, 0.9 brown

ANALYSIS:

Since brown is dominant, the 10 green individuals must be homozygous and will contribute 20 green genes to the gene pool. If the 10 brown individuals are all homozygous they will contribute 20 brown genes and the gene frequency for each will be 0.50.

If the 10 brown are all heterozygous, then they will contribute 10 brown and 10 green genes. The green individuals will still contribute 20 green genes. The frequency of the brown gene will then be $10/40 = 0.25$, and of the green gene, $30/40 = 0.75$. Thus, the gene frequency for brown could be between 0.25 and 0.50, and for green between 0.50 and 0.75. Choice 3 is the only one which falls within this range.

EXAMPLE 2

In nature mice are subject to a bacterial disease to which heterozygous individuals are resistant. The homozygous dominant individuals are not resistant to the disease while homozygous recessive individuals die at a very young age due to the products of the recessive genes. If a population of these mice is placed in the laboratory under sterile conditions what do you predict will happen to the frequency of the recessive gene?

    1) The frequency of the gene will remain constant because of random mating.
    2) The frequency will decrease because of the mortality of the homozygous recessive individuals.
    3) The frequency will decrease because of the loss of the advantage enjoyed by heterozygous individuals over homozygous dominant individuals.
    4) Both 2 and 3 are reasonable predictions.
    5) The frequency will increase due to the unnatural laboratory conditions.

ANALYSIS:

In the natural environment the homozygous recessive individuals die young. If it is assumed that they die before reaching reproductive age, then they will make no contribution to the gene pool in the next generation. The homozygous dominant individuals will be selected against since they are subject to death by disease.

In a sterile environment selection will no longer operate against the homozygous dominant individuals. The homozygous recessive individuals will still die without reproducing and the resistant heterozygous individuals will be unaffected.

1. Random mating could occur in both the natural and the sterile environment but gene frequencies may change because different selection pressures are acting in the different environments.

56

2. The mortality of the homozygous recessive individuals should be the same in both environments since it is due to factors independent of the environmental conditions under consideration. The heterozygotes, however, will continue to transmit the recessive gene to subsequent generations at the same rate. Since these two factors will be unchanged, they alone do not predict a change in gene frequency.

3. The advantage of the heterozygotes lies in the selection against those homozygous for the dominant gene. This selection will not operate in a sterile environment and the homozygotes will reproduce faster than they do in the wild. This will result in an increase in the frequency of the dominant gene. This would occur even if the homozygous recessive condition was not lethal. An increase in frequency of one allelle necessarily results in a decrease of frequency of the other.

4. Reject because choice 2 has been rejected.

5. Reject because the unnatural conditions lead to the prediction of a decrease in frequency. See the argument for choice 3.

1. Consider a population with the following structure:

Genotype:	AA	AB	BB
Number of Individuals:	60	20	20

Which of the following most accurately represents the gene frequencies in this population?

Frequency

	A	B
1)	0.6	0.2
2)	0.7	0.3
3)	0.8	0.4
4)	1.2	0.4
5)	None of the above.	

2. In a population three genotypes were present, AA, Aa, aa each with equal reproductive success. If the genotypic frequencies are:

$$AA = .04 \qquad Aa = .32 \qquad aa = .64$$

What will the gene frequencies of A and a be in the second generation?

1) A = .6 a = .4
2) A = .5 a = .5
3) A = .4 a = .6
4) A = .3 a = .7
5) A = .2 a = .8

3. A biologist had 31 dark mice and 31 light mice. She bred these mice so that a dark and a light mouse were always mated; all offspring were dark. Which of the following would be the gene frequencies for the dark gene (D) and for the light gene (L) in the parents and offspring?

	Parents		Offspring	
	D	L	D	L
1)	1.0	1.0	2.0	0
2)	0.5	0.6	1.0	0
3)	1.0	1.0	1.0	1.0
4)	0.5	0.5	0.5	0.5
5)	Cannot be determined from the information given.			

4. A red carnation and a white carnation are crossed and produce 500 seeds. When these seeds are planted all the resulting plants have pink flowers. What is the gene frequency of the red and white genes in this population of plants with pink flowers?

1) Red = .1, white = .9
2) Red = .5, white = .1
3) Red = .8, white = .2
4) Red = .5, white = .5
5) Red = 1.0, white = 1.0

5. There were two varieties of a certain species of fish found in the Monongahela River. There were two genes present for spots; SS genotype had many spots, NN had no spots and SN had a few spots. A sample of these fish contained 312 many spotted, 312 not spotted, and 745 few spotted. Which of the following best describes the gene frequencies in the sample?

1) S greater than N
2) N greater than S

3) S equal to N
4) S = .45, N = .55
5) In the Mon River — who knows?

6. A student studying a population of butterflies wanted to determine the gene frequencies for wing color. She knew that in this butterfly wing color is exhibited as either yellow or black and that black is dominant. She also knew that half the butterflies in this population had yellow wings and half had black wings. Which of the following possibilities would she most likely reject?

1) There are more yellow genes than black genes.
2) There are more black genes than yellow genes.
3) There are twice as many yellow genes as black genes.
4) There are equal numbers of black and yellow genes.
5) There are 3 times as many yellow genes as black genes.

7. Suppose a hog breeder discovered a recessive gene for straight tail in hogs. After some years of selective breeding he owned the following population of hogs:

Straight tailed   25
Curly tailed   75

Which of the following is the most likely frequency of the gene for straight tails in this population?

1) 0.125
2) 0.225
3) 0.625
4) 0.725
5) 0.75

8. If the hog breeder in Question 7 acquired 10 new hogs heterozygous for curly tail and added them to his population which of the following would most likely be the new frequency of the gene for curly tail in the population?

1) Greater than before.
2) The same as before.
3) Less than before.
4) 1 and 3 are equally likely.
5) 1, 2 and 3 are equally likely.

9. You are studying a sex-linked, recessive gene, F. You examine the following population:

# of individuals	Sex	Phenotype
20	male	F
20	male	not F
10	female	F
10	female	not F (homozygous)
10	female	not F (heterozygous)

What is the frequency of gene F in this population?

1) 0.100
2) 0.250
3) 0.357
4) 0.500
5) 0.700

10. Suppose there is a population of organisms which exhibit two expressions, O and P, of the same characteristic. It is known that P is dominant over O. Which of the following would decrease the gene frequency of P most rapidly?

1) If O becomes dominant over P.
2) If all females prefer to mate with homozygous recessive males.
3) If all females prefer to mate with males exhibiting the dominant trait.
4) If females mate indiscriminately with males regardless of O/P expression.
5) If all females preferred to mate with homozygous dominant males.

11.   Suppose that a lake in Nebraska contained 27,342,824 individuals of a certain hypothetical organism. There were two genes, designated P and G, respectively, which determined color in this organism. Genotype PP was pink; GG was green and PG was white. Suppose a tourist who was eight years old caught 40 white individuals in a jar. What are the most likely gene frequencies in the lake?

1) P greater than G.
2) P equal to G.
3) P less than G.
4) P = 0.25, G = 0.75.
5) Can't tell, but if he had caught 40 more, we could tell.

12.   Suppose the boy released his 40 individuals into a lake in West Virginia where this organism did not formerly live but had no natural enemies. Assume 75% of them survived, bred randomly and reproduced successfully. After 10 generations what would be the most likely gene frequencies in the West Virginian lake?

1) P greater than G.
2) P equal to G.
3) P less than G.
4) Can't tell.
5) Can't tell, but if they had all survived, we could tell.

13.   Suppose that, after these 10 generations, this organism had reproduced so successfully that it was a pest. In order to control them the West Virginia Department of Natural Resources brought 10 fish from Nebraska which were voracious feeders on these organisms and put them into the lake. They survived and bred randomly but, unfortunately, the original 10 fish were homozygous for a trait which made them able to see only white objects. After 10 more generations what would be the most likely gene frequencies in the lake?

1) P greater than G.
2) P equal to G.
3) P less than G.
4) Can't tell.
5) Can't tell, but if the fish could see only pink objects, we could.

14.   Suppose a change occurred in the water composition in the lake which made eggs from green females twice as likely to be fertilized and develop into adults. The likelihood of fertilization by sperm from green, white or pink males remained equal. After 10 more generations what would be the most likely gene frequencies in the lake?

1) P greater than G.
2) P equal to G.
3) P less than G.
4) Can't tell.
5) Can't tell, but if we knew the genotypes of sperm from white males, we could tell.

15.   In a population of 100 mice two genes, A and B, are found. A is recessive and its expression leads to the death of a mouse at a very early age. If the original gene frequencies were 50% for gene A and 50% for gene B what could you most confidently conclude about this population in 20 generations? Assume random mating and no migration.

1) The frequency of gene A will be greater in reproductively mature female mice than in reproductively mature male mice.
2) The frequency of gene A will be unchanged in the population.
3) The frequency of gene B will decrease in reproductively mature females.
4) All reproductively mature mice will be homozygous for gene B.
5) No reproductively mature mice will be homozygous for gene A.

16.   Suppose in the previous question it was assumed that the expression of gene A was a recessive sex-linked trait. Again assuming random mating and no migration which of the following could you most confidently conclude about this population in 20 generations?

1) All reproductively mature mice will be homozygous for gene A.
2) All reproductively mature mice will be homozygous for gene B.
3) The frequency of gene B will be unchanged.
4) The frequency of gene A will be greater in reproductively mature female mice than in reproductively mature male mice.
5) The frequency of gene A will be unchanged.

17.   A biologist exposed a mosquito population to a moderate concentration of DDT; approximately 2% of these mosquitos survived. These surviving mosquitos were allowed to interbreed and produce offspring. When the offspring were exposed to DDT, approximately 94% survived. Which of the following is the best conclusion you could draw from these observations alone?

1) DDT resistance is a beneficial trait for mosquitos in the natural environment.
2) Exposure to DDT induced a mutation to DDT resistance which was inherited by the offspring.
3) A mutation for DDT resistance is a recessive gene and is expressed in homozygous individuals.
4) Exposure to DDT allows only those mosquitos to survive which already have genetic information for DDT resistance.
5) Choices 2 and 4 are equally likely.

18.   A bacteriologist wanted to repeat the velvet pad replica plating experiment with a newly discovered bacteria. He assumed that in these bacteria exposure to streptomycin can induce random mutations for streptomycin resistance. Which of the following would he most likely predict after making this assumption?

1) The location of a colony on a replicate streptomycin plate will indicate whether the colony is resistant to streptomycin.
2) When replicates are made with the velvet pad on streptomycin plates, resistant colonies will appear at different positions on each plate.
3) When a colony has developed streptomycin resistance, this resistance will be passed along to the offspring.
4) If the velvet pad is used to make replicate plates containing NO streptomycin, few of the colonies will survive.
5) Streptomycin resistance is dominant.

19.   A current biology textbook states that "most new mutations are recessive," but offers no evidence to support the statement. Which of the following assumptions would be the most important in an argument for the book's statement?

1) Most individuals are heterozygous for most genes.
2) Most individuals are homozygous for most genes.
3) Most existing genes are recessive.

4) Most existing genes code for active enzymes.
5) Mutations are rare events.

20. For several years spider mites were successfully controlled with a pesticide called parathion, but now in many areas most mites are resistant to this chemical. This resistance is associated with an increase in the frequency of a dominant gene for resistance to parathion. Which of the following best explains the increase in the parathion-resistance gene?

1) The mites adapted to parathion in order to keep from becoming extinct.
2) Mites which had genes for resistance to parathion left more surviving offspring than those which lacked the gene.
3) A few of the spider mites acquired a mutation for resistance to parathion; then the gene spread through the population because it is dominant.
4) The gene increased to its present high frequency because mutations for resistance to parathion occurred following use of the parathion.
5) 1, 2 and 4 are all good explanations for change in frequency of the gene.

21. In any group of adult human patients treated with penicillin, 1-5% will exhibit some form of allergic response. Allergic responses result from the interaction of the penicillin molecule with a specific protein molecule in the patient's blood. A student suggested that penicillin treatment caused a mutation which led to production of the specific protein molecule in a few individuals. Which of the following is the best alternative interpretation?

1) The gene for allergic responses is dominant.
2) Penicillin treatment exerts a strong selective pressure against the gene for allergic responses.
3) Penicillin treatment exerts a strong selective pressure for the gene for allergic responses.
4) The gene which carries information for the specific protein molecule is present in a few individuals before treatment with penicillin.
5) Allergic responses lead to penicillin resistance.

22. A Biology student placed 50 red-eyed flies and 50 white-eyed flies in a sealed chamber and determined the frequency of the gene for red eyes in each succeeding generation. The frequency fell rapidly for the first ten generations, then stabilized at approximately 0.1 for ten more generations. Which of the following statements would best explain her results?

1) Red-eyed females prefer to mate with white-eyed males.
2) White-eyed females mate only with white-eyed males.
3) Red-eyed male offspring are sterile.
4) Red-eyed flies carry a recessive lethal gene.
5) All the above.

23. In a population of humans, if the frequency of the gene for hemophilia was 1 in 25,000, what would most likely happen to the gene in future generations?

1) Since it is recessive, it should drop out of the population before long.
2) One hemophiliac in 25,000 would have trouble meeting another hemophiliac, hence they would not reproduce, and the gene would soon be lost.
3) Even a small chance of meeting might produce some hemophiliac offspring, and since mutation might add to the recessive gene numbers, the gene would continue in very low numbers in the population.
4) All the above are correct.
5) None of the above is correct.

24. A botanist was studying three species of grass plants. Species A had 14 chromosomes, species

B had 14 chromosomes, and species C had 21 chromosomes. He found that species A and C had 7 chromosomes that were identical and that species B and C had 14 chromosomes that were identical. Which of the following is the best explanation of the data?

1) Gametes containing a haploid number of chromosomes from species B were fertilized by a gamete containing the diploid number of chromosomes from species A.
2) All three plants reproduce by an asexual and then a sexual generation of haploid and then diploid plants.
3) Species C has a haploid number of 10 chromosomes in male gametes and 11 chromosomes in female gametes.
4) Species C is a hybrid grass plant of species A and species B.
5) Species C is only found in an area of overlap of species A and species B.

25. A biologist collected plants similar in appearance from different parts of Texas. The range of plants from any one site did not overlap the range of plants from any other site. He later tried interbreeding the plants obtained from three different sites. The percentages of fertile offspring resulting from each cross are shown below.

	Site 1	Site 2	Site 3
Site 1	100%	25%	14%
Site 2	25%	100%	3%
Site 3	14%	3%	100%

Which of the following could you most confidently infer about these plants?

1) In the natural environment where these plants grew they commonly would interbreed to produce seed.
2) Hybrids produced by crossing plants from different sites are less able to survive than are the other plants.
3) Site 2 and site 3 have considerably different environments.
4) Most of the plants at each site are genetically homozygous for most traits.
5) At some time in the past plants of this type had composed one continuous interbreeding population.

26. A botanist had collected some plants from California and from Texas. When comparing these plants they were identical in appearance, yet when he attempted to interbreed these plants NO seeds were produced. Which of the following would most likely account for these observations?

1) Plants from California and from Texas are widely separated by distance and cannot interbreed.
2) Whether a plant does or does not produce seed is not as important as the number of seed which germinate and produce offspring.
3) Genetic differences could be present which prevent fertilization but have no effect on the plant's appearance.
4) If plants are identical in their outward appearance, it would be expected that they could interbreed and produce fertile seeds.
5) Choices 2 and 4 are equally likely.

27. One kind of frog lives in the area stretching from Georgia to Maine. Frogs in this area apparently form a continuous population with no obvious barriers. It has been noticed however, that although frogs can interbreed with frogs in nearby areas, frogs from the extremes of the range (i.e., frogs from Georgia and frogs from Maine) will NOT interbreed. This example would cause you to most seriously question which of the following?

1) Gene frequencies in a continuous population will remain constant unless there is migration of individuals with different gene frequencies.

2) Mutations may occur in a population which can change the genotype of some members of the population.
3) Genes can move or flow from one part of a continuous gene pool to all other parts.
4) Random breeding will occur in a population unless there are mechanisms which prevent some individuals from breeding with others.
5) In a continuous population stretching over a large area, environmental factors may be different in different parts of the world.

28. Which of the following would best test the hypothesis that two similar, but isolated, populations of squirrels were different species?

1) Examine the populations to make sure the differences are present in all squirrels.
2) Allow the populations to interbreed and see if offspring are viable.
3) Compare the rate of reproduction in both populations.
4) Compare the environmental conditions.
5) Compare the rate of mutations in both populations.

29. Suppose you observe two populations of plants which resemble each other very closely and which occupy ranges that do not overlap. There are two hypotheses:

1. The two populations were once one.
2. The two populations were originally different but became similar.

You observe that individuals from the two populations can interbreed and 10% of the hybrids are fertile. Your biology instructor claims that this evidence favors hypothesis 1. Which of the following assumptions is he most likely making?

A. In order to produce fertile offspring, two organisms must have many genes in common.
B. The same mutations are not likely to occur in two different populations.

1) A only.
2) B only.
3) Both A and B.
4) Either A or B, but not both.
5) Neither A nor B.

30. When dogs were introduced to Australia, the more primitive hunting mammals which had formerly been present there soon became extinct. However, even with the more efficient predation, the prey species survived. One new Biology textbook says this is because, "as the prey begin to be killed off, the predators find themselves with less food, and so their own numbers fall off due to starvation." Which of the following assumptions are the authors most likely making?

1) These dogs cannot switch to another species for food.
2) These dogs preyed upon the original hunting mammals.
3) These dogs reproduce equally well on any diet.
4) The prey animals compensate for their smaller population size by reproducing faster.
5) When there are fewer dogs in a hunting pack, the prey animals are more likely to escape.

31. Sometimes similar kinds of plants will interbreed to produce offspring which exhibit "hybrid vigor". These hybrids grow taller, are physically stronger, more resistant to disease, and have other characteristics which make them superior to the parent plants. One would think that these hybrid plants would become much more common than either of the parent plants. However, this seldom occurs. Which of the following would be the best explanation?

1) The hybrid plants cannot produce offspring.
2) Mutation may occur more frequently in the parent plants.

64

3) Hybrids have some characteristics of one parent and some characteristics of the other.
4) The hybrid plants reproduce asexually.
5) For genetic change to occur in two similar groups of organisms, they must be separated by some type of barrier.

32. Two closely related species may inhabit the same area and be capable of interbreeding but rarely, if ever, interbreed in the wild. Interbreeding may be prevented by a variety of mechanisms such as different breeding seasons. Which of the following would most favor the development of mechanisms to prevent interbreeding between species?

1) Hybrids produced by interbreeding are sterile.
2) One species is much less successful when competing with the other for mates.
3) One species has a higher frequency of dominant genes.
4) The two species are adapted to different environments.
5) Both 3 and 4 are equally likely as the best choice.

33. Horses and donkeys are regarded as separate species. Crossing a horse with a donkey produces a healthy and stubborn but sterile animal called a mule. Suppose a horse gene undergoes a mutation such that the cells of any individual with that mutated gene can be easily recognized. In which of the following is the gene LEAST likely to occur?

1) A zygote produced by a horse.
2) A zygote produced by a donkey.
3) The skin cells of a mule.
4) The descendant of a Shetland pony.
5) The gamete of a donkey.

34. If horses and donkeys were left together for a sufficient number of generations which of the following genes would be most likely to increase in frequency?

1) A gene that increases the frequency of horse-donkey hybrids.
2) A gene that makes the horses avoid donkeys.
3) A gene that makes horses and donkeys look and sound more alike.
4) A gene that maintains the sterility of mules.
5) A gene that increases the resistance of mules to disease.

35. Consider two similar birds, A and B, which live all along the northern United States from Maine to Washington. In Washington, only A is found and it breeds in early July. In Maine, only B is found and it also breeds in early July. In Minnesota both birds are found, A breeds in late June and B breeds in late July. Which of the following predictions is most likely to be the case?

1) A and B can interbreed and the hybrids are fertile.
2) A and B can interbreed but the hybrids are sterile.
3) A and B cannot interbreed.
4) A and B both build nests in trees.
5) A and B both eat crenelated bark beetles.

The following description is used in questions 36, 37 and 38.

A biologist was studying the lizards present on an island in the Atlantic Ocean. From previous work he knew that all the lizards on the island could interbreed. In his latest experiment he investigated the expression of three traits in lizards on various parts of the island:

Trait 1 — expressed as A or B.
Trait 2 — expressed as W or Y.
Trait 3 — expressed as K or J.

From three sites on the island he collected lizards, and recorded the various phenotypes and recorded the percentage of lizards with traits shown below:

Phenotypes	Site 1	Site 2	Site 3
AWJ	0%	3%	90%
BWJ	27%	9%	1%
AYK	0%	84%	0%
BWK	0%	4%	0%

36.   Which of the following could you most likely infer?

1)   The environmental conditions at sites 2 and 3 are considerably different.
2)   Lizards can sometimes travel from one site to another.
3)   Lizards with phenotypes AWJ possess a combination of traits which favor their survival at site 3.
4)   If lizards with phenotype AYK are introduced into an environment different from the three examined here, they should survive much better.
5)   Choices 1 and 3 are equally likely.

37.   A biologist predicted that if lizards with the phenotype BYK were released at site 2, they would not survive very well. Which of the following is the most persuasive argument supporting his prediction?

1)   Lizards with BYK phenotypes compose only 4% of the population at site 2, thus lizards with a BYK phenotype should not do well.
2)   There is no information in the data to indicate that BYK lizards will survive at site 2.
3)   The biologist apparently did not look for BYK lizards at any of the sites.
4)   Sexual breeding between lizards at site 2 should produce some BYK lizards, yet none are present.
5)   Most of the lizards at site 2 have the phenotype AYK.

38.   After examining the data shown for site 3, a student suggested that A is dominant over B. Which of the following would be the most serious criticism you could make of his suggestion?

1)   The problem states that trait 1 is expressed as A or B, but there is no intermediate expression.
2)   Expression A could be recessive but be a more advantageous expression of trait 1 at site 3.
3)   At site 3 traits 1 and 3 are expressed only as W and J.
4)   If A is recessive, all the AWJ organisms at site 3 would be homozygous for A.
5)   If B were the dominant expression of trait 1, it should appear more frequently at site 3.

39.   In a marine fish it is known that tail length, fin shape and body color are all genetically coded traits. Further, expression of tail length may be either long or short, fin shape may be either round or pointed, and color may be blue, green or yellow. A biologist who was interested in the genetics of these characteristics crossed a trihomozygous long-tailed, round-finned, blue male with a trihomozygous short-tailed, pointed-finned, yellow female. The resulting offspring were all long-tailed round-finned green individuals. The biologist then crossed a female offspring with a male which was short-tailed pointed-finned and yellow. The offspring from the second cross were as follows: one-half were long-tailed, round-finned green and one-half were short-tailed, pointed-finned yellow. Which of the following is the best conclusion the biologist could draw from his experiment?

1)   These three traits are linked on the sex chromosomes.
2)   Tail length and fin shape are linked but color is not.
3)   These three traits are linked on one pair of non-sex chromosomes.
4)   Color and tail length are linked on one pair of non-sex chromosomes.

5) There is no linkage but there is dominance.

40. Another biologist questioned the findings of the biologist in the previous question. She repeated the second cross, just as the biologist did before, and observed the following offspring:

Frequency (%)	Tail length	Fin shape	Color
40%	long	round	green
4	long	round	yellow
4	long	pointed	green
4	long	pointed	yellow
3	short	round	green
3	short	round	yellow
3	short	pointed	green
39	short	pointed	yellow

With this additional information, which of the following is the best conclusion to be drawn?

1) Because there are eight phenotypes, these traits are not linked.
2) The differences in the results of the two crosses are due to dominance effects.
3) The data are unclear because two of the offspring should have been crossed.
4) The data are consistent with the hypothesis that the genes are linked.
5) The data are contradictory and the experiment should be repeated more carefully.

41. A population of the fish in the previous problem consisted of the following individuals:

Males	Blue	33
	Green	58
	Yellow	77
Females	Blue	67
	Green	142
	Yellow	23

If for this problem you assume that tail length and color are linked in these fish, what was the frequency of the gene for short tails in this population?

1) 0.23
2) 0.33
3) 0.50
4) 0.67
5) 0.77

42. If green males of the fish in the previous problem could not see green objects, what would most likely happen to the frequency of the gene for long tails?

1) Increase rapidly to 1.0.
2) Increase to some steady level.
3) Stay the same.
4) Decrease to some steady level.
5) Decrease rapidly to 0.

43. Now assume that the genes for color and for fin shape are linked in the fish in the previous problem. suppose a predator is introduced which catches round-finned individuals more often than pointed-finned individuals. Which of the following is most likely to occur?

1) The gene for pointed-fins will become dominant.
2) The gene for green will decrease in frequency.
3) The number of yellow individuals will increase.

4) The gene for long tails will increase in frequency.
5) The gene for round fins will become recessive.

RESPIRATION, PHOTOSYNTHESIS, AND FERMENTATION

The energy contained in organic material can be estimated conveniently by burning a sample in a bomb calorimeter. Energy is held within the chemical bonds of the molecules comprising the material; much of this energy is released when the bonds are broken during oxidation — i.e., burning. In living organisms the same overall process occurs except that bonds are broken in the step-wise process of respiration. The primary source of energy is carbohydrates. Some of the energy in the chemical bonds is transferred to other chemical bonds, primarily in the conversion of adenosine diphosphate (ADP) to adenosine triphosphate (ATP) and some is released as heat. The end products of the reaction are carbon dioxide and water. Some of the energy in ATP can be released by enzymes and utilized in energy-requiring processes such as muscle contraction and active transport. When ATP is used in this way, ADP is regenerated.

Many organisms break down carbohydrates only partially to produce alcohol and carbon dioxide as end products, with no oxygen utilization. This process, termed fermentation, releases much less energy than respiration and consequently forms much less ATP.

Although most living organisms simply transfer chemical bond energy from one type of molecule to another, many plants can convert radiant energy to chemical bond energy in the process of photosynthesis. Light energy is absorbed by photosynthetic pigments and is transferred to chemical bonds when sugars are formed from carbon dioxide and water obtained from the external environment. Free oxygen is also a product.

EXAMPLE 1

When a gram of glucose is burned in the bomb calorimeter, 3833 calories of energy are released. Presumably, this represents the energy available to an organism when it breaks down glucose. A student commented that 3833 calories does NOT accurately represent the energy available to an organism since ATP is also produced by the organism. The student further stated that glucose and ATP should be burned in the calorimeter to measure the energy available to the organism. Which of the following would be the best criticism of the student's comments?

1) Rather than ATP, ADP should be burned along with the glucose.
2) ATP transfers energy to other locations.
3) ATP gets its energy from the breakdown of glucose.
4) Enzymes are necessary to convert ADP to ATP when glucose is broken down.
5) The energy in ATP is released at a different time than the energy in glucose.

ANALYSIS:

Living cells release chemical bond energy from glucose in a process resembling the reactions in the calorimeter. Some of the released energy is conserved in a useful form by converting ADP to ATP. Thus the useable energy in ATP comes from glucose. Burning ATP and glucose in the calorimeter would release and measure this energy twice.

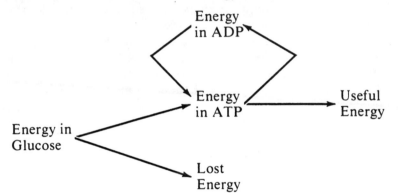

1. This worsens the error of burning ATP. The energy in ADP is not available to the organism at all, since this compound is recycled.

2. This is an acceptable statement but it does not criticize the comments in the problem. The energy content of ATP and the source of its energy are the same whether it moves or not.

3. This is the criticism detailed above. The student proposes to measure the same energy twice.

4. This also is an acceptable statement, but not a criticism. Energy content and source of energy are not changed by the reaction mechanism.

5. This statement could be viewed as supporting rather than criticizing the student. Since the energy is released at two different times, it could represent two separate sources which should be added together. However, the two molecules are related sequentially rather than in parallel.

EXAMPLE 2

Prior to 1930 scientists thought that in photosynthesis light energy separated the carbon and oxygen in CO_2. The carbon then combined with water to form starch and the oxygen from CO_2 was released as oxygen gas. Which of the following would provide evidence contradicting this view?

1) Expose a lighted plant to CO_2 with radioactive carbon.
2) Increase the illumination of the plant.
3) Supply the plant with more water vapor in the surrounding air.
4) Measure the total amount of starch and oxygen produced by the plant over a given time period.
5) Demonstrate oxygen evolution from plant cells when no CO_2 is present.

ANALYSIS:

Evidence contradicting this view must indicate a source of O_2 other than CO_2.

1. This would indicate the fate of carbon in CO_2 but would give no information about O_2.

2. Increased illumination would be expected to increase O_2 production but would provide no evidence about the O_2 source.

3. Providing more water vapor would most likely have no effect on photosynthesis, much less any effect which would indicate the O_2 source.

4. Measuring starch and O_2 produced would provide information on the rate of photosynthesis only.

5. If a photosynthetic plant can produce O_2 when no CO_2 is present, then the O_2 source must be some substance other than CO_2.

1. The following observations of food eaten and dung produced by a 200 kg cow were made over a two week period.

	FOOD	DUNG
WEIGHT	15 Kg/day	14.7 Kg/day
ENERGY	60 Kcal/day	30 Kcal/day

Which of the following is the best interpretation that can be made from these data?

1) There is more total energy in dung produced than in food eaten.
2) The cow gets all of its energy from sunlight.
3) More energy per unit weight is found in the dung than in the food.
4) The cow somehow removes half of the energy that is in the food eaten.
5) The cow gains 0.3 kg of wt. per day.

2. A student examined the data shown below on the total energy in food consumed and dung produced each day by a cow.

FOOD	DUNG
(Kcal/day)	(Kcal/day)
54	24

He then concluded that the weight of food consumed each day is about twice the weight of dung produced. Which of the following assumptions is most likely being made by the student?

1) The energy contents per unit weight of food and dung are equal.
2) Energy contents of food and dung can be measured in the same way.
3) The weight of food taken in each day can be used to determine the amount of dung produced.
4) Dung is produced almost entirely from food consumed.
5) Energy is not related to weight in a systematic way.

3. An animal physiologist collected the following data on food consumed and dung produced by a calf over a 60 day period.

Body weight day 1 = 100 kg dry weight.
Body weight day 61 = 115 kg dry weight.
Food consumed in the two months = 480 kg dry weight.
Dung produced in the two months = 450 kg dry weight.

The physiologist noticed that the difference between food consumed and dung produced was 30 kg but that the calf only gained 15 kg. He could not understand the discrepancy and concluded that his measurements must be in error. Which of the following is the best interpretation of the physiologist's data?

1) The data are acceptable and the calf converted 15 kg of food into energy.
2) The data are acceptable and some of the calf's weight gain came from oxygen inhaled.
3) The data are acceptable and the calf lost matter as carbon dioxide exhaled.
4) The data are acceptable and not all the food consumed by the calf was digested.
5) Explanations 1-4 are all unreasonable and the data must be rejected.

4. When a plant is germinated and grown for a few days entirely in the dark it is found that the plant contains a significant amount of energy. A student hypothesized that this energy was originally present in the seed. To test her hypothesis she gathered the following data:

Average energy in seed = 5.76 KCal/gram
Average energy in plant = 4.57 KCal/gram

She recognized that these data were insufficient for her to draw a conclusion. What other data would she most likely need?

1) Average amount of energy expended by a plant when growing to the size used in the experiment.
2) Average amount of energy present in the soil used to grow these plants.
3) Average weight of a single seed.
4) Average weight of a single plant.
5) Both 3 and 4 are needed.

5. A student grew two sets of corn seeds. After 14 days she recorded the following data:

	Fresh wt. (gm)	Dry wt. (gm)	Kcal/gm dry wt.
Set A (16 hrs. of light/day)	14.0	6.1	1.0
Set B (8 hrs. of light/day)	14.2	3.2	1.0

Which of the following is a valid interpretation of these data?

1) Set A contains the same amount of energy as Set B.
2) Set A contains more energy than Set B.
3) Set A contains less energy than Set B.
4) Set B was albino.
5) Plants grown for the same number of days contain the same amount of energy regardless of light conditions.

6. A zoologist wanted to estimate the amount of vegetation eaten by rabbits in a specified area. He collected all the fresh rabbit dung in the area. Assuming the rabbits eat only plants, which of the following would probably give him the most accurate estimate of the mass of vegetation eaten by rabbits?

1) The total KCal in the rabbit dung divided by the average KCal/g in plant material.
2) The total KCal in the sunlight falling on the area divided by the average KCal/g in plant material.
3) Twice the total KCal in the rabbit dung.
4) Estimate the total KCal in the vegetation growing in the area.
5) The KCal/gram in rabbit flesh times the KCal/gram in plant material.

7. The zoologist in Question 6 would most likely underestimate the amount of vegetation eaten by the rabbits if he does not know which of the following?

1) The amount of energy in sunlight falling on the area.
2) The amount of energy rabbits extract from the food they eat.
3) The average energy in KCal/gram of plant material growing in the area.
4) How many rabbits are present in the area where the measurements were made.
5) The total energy in KCal of the vegetation growing in the area.

8. Which of the following would produce the greatest energy when burned in the bomb calorimeter?

1) 10 grams of glucose
2) 20 grams of CO_2
3) 10 grams of starch
4) 20 grams of CO_2 plus 5 grams of glucose.
5) 20 grams of CO_2 plus 5 grams of starch.

9. A scientist broke down the chemical bonds between the sugars in a starch molecule and

measured the amount of energy released. He then burned the individual sugars in a bomb calorimeter and found that more energy was released in the calorimeter than when he separated the sugars. Which of the following is the best conclusion that can be drawn from these data?

1) Starch has less energy because it is a larger molecule than sugar.
2) There are more chemical bonds within sugars than between sugars in starch.
3) There are fewer chemical bonds within sugars than between sugars in starch.
4) Starch can absorb light energy and therefore contains more energy.
5) Starch weighs more than sugar and therefore contains more energy.

10. A biologist studied samples of millet grain and sunflower seeds in the bomb calorimeter. He observed that millet grain released more kilocalories per gram than sunflower seeds. He assumed that food serves as the only source of energy for an animal and concluded that millet grain was better food for his pet birds than sunflower seeds. Which of the following assumptions was he also making?

1) The reactions in the calorimeter released all the chemical bond energy in the samples.
2) Millet grain costs less per kilocalorie than sunflower seeds.
3) The energy released in the calorimeter comes from the same chemical bonds as the energy released inside the birds.
4) Millet grain has the same chemical composition as sunflower seeds.
5) The reactions in the calorimeter are exactly the same in all respects as the reactions inside the birds' cells.

11. The biologist in Question 10 carefully measured the weight of the sunflower seeds placed in the calorimeter. He verified that the seeds were completely burned to carbon dioxide, water and a little mineral ash. He found that the weight of these three products was greater than the weight of the seeds. How would you interpret this observation?

1) This result is expected when organic material is burned.
2) There must have been an error in the weighing.
3) This leads to the conclusion that mass is converted into energy.
4) The reactions in the calorimeter release all the chemical bond energy in the samples.
5) It was an error to neglect to consider the water content of the seeds.

12. A mouse was found to utilize 850 cubic millimeters of oxygen per gram of body weight per hour. An elephant was found to utilize only 100 cubic millimeters of oxygen per gram of body weight per hour. A student stated that the elephant was going to die soon because his cells were not using enough oxygen to sustain adequate cellular respiration. Which of the following is the best criticism of the student's statement?

1) Elephants might need more O_2 per gram of body weight per hour than mice to maintain cellular respiration because elephants have more cells than mice.
2) Elephants might have a higher metabolic rate than mice so they would require less O_2 per gram of body weight per hour than mice.
3) Elephants might have a lower metabolic rate than mice so they would require more O_2 per gram of body weight per hour than mice.
4) Elephants might have a lower metabolic rate than mice so they would require less O_2 per gram of body weight per hour than mice.
5) The student's statement cannot be criticized.

13. A + B $\longrightarrow$ C $\longrightarrow$ D $\longrightarrow$ E

A scientist advances the hypothesis that these reactions are part of a cycle. She thinks that further reactions will lead to the production of A again. Which of the following observations would give the strongest support to her hypothesis?

1) Blocking the conversion of C to D leads to an increase in the concentration of B.
2) Blocking the conversion of C to D leads to a decrease in the concentration of E.
3) Molecules of A containing radioactive atoms are added and radioactive molecules of A are found in the system later.
4) Radioactive molecules of B are added and radioactive molecules of compound X are later found.
5) Radioactive molecules of C are added and radioactive molecules of A are later found.

14. In his autobiography, Ben Franklin tells of arguing with a fellow printing apprentice "... that the bodily strength afforded by beer could only be in proportion to the grain or flour of the barley dissolved in water of which it was made, that there was more flour in a pennysworth of bread, and therefore if he would eat that with a pint of water, it would give him more strength than a quart of beer." If we interpret Ben's use of "strength" as energy, which of the following statement is most accurate today?

1) Ben's conclusion is wrong because alcohol contains more energy of combustion per gram than does starch (flour).
2) Ben's conclusion is correct because beer would contain the same amount of energy as in the grain or flour used in making it.
3) Ben's conclusion is wrong because beer contains more energy than the materials used in making it.
4) Ben's conclusion was more correct than he supposed because the beer would contain less energy than the materials used in making it.
5) Ben's conclusion remains in question because there are no samples of 18th century English beer available for calorimetric analysis.

15. A colony of a particular microorganism was grown in a closed container in a medium containing glucose and amino acids and with a surrounding atmosphere consisting of pure CO_2. The colony thrived and reproduced. Which of the following must have been occurring in the cells of the microorganism?

1) Photosynthesis.
2) Anaerobic breakdown of glucose.
3) Aerobic respiration.
4) Photosynthesis and anaerobic breakdown of glucose.
5) Photosynthesis and aerobic respiration.

16. Each of the ten flasks containing an appropriate culture medium is inoculated with equal numbers of an organism that can grow and reproduce under aerobic and anaerobic conditions. After a period of time, the biomass in each flask is measured. The following are the results:

Flask Number:	1	2	3	4	5	6	7	8	9	10
Biomass:	0.41	0.40	0.41	0.21	0.22	0.21	0.19	0.22	0.20	0.21

Which of the following is the most likely explanation for these results?

1) The cultures in flasks 1, 2 and 3 were grown under aerobic conditions while the others were grown under anaerobic conditions.
2) The cultures in flasks 1, 2 and 3 were grown under anaerobic conditions while the others were grown under aerobic conditions.
3) All cultures were grown under anaerobic conditions but only those in flasks 1, 2, and 3 survived to the end of the observation period.
4) Three of the cultures were grown in the dark while the others were kept in the light.

5) Flasks 7, 9, and 10 were refrigerated for the first half of the observation period while the others were maintained at room temperatures; thereafter all were incubated at 30° C.

17. When a microbiologist grew some yeast under aerobic conditions he found that 24 units of oxygen were used and 24 units of carbon dioxide were produced. He then grew a second culture of yeast under aerobic conditions for a time then made it anaerobic for a time. Which of the following data would you most likely predict?

1) O_2 used = 36 units, CO_2 produced = 24.
2) O_2 used = 36 units, CO_2 produced = 36.
3) O_2 used = 12 units, CO_2 produced = 24.
4) O_2 used = 27.4 units, CO_2 produced = 19.5 units.
5) Either 1 or 2 would be equally likely as the best prediction.

18. A student sets up an apparatus to produce alcohol. He includes yeast cells, water and glucose in excess and maintains it in constant darkness. For 2 weeks he notices an increase in yeast biomass and CO_2 produced. After 4 weeks he observes a decrease in biomass and very little production of CO_2. Glucose is still present. Which of the following conclusions would be strongest?

1) The yeast cells have totally depleted their energy source.
2) Since ATP is formed during anaerobic respiration, this ATP has poisoned the yeast cells.
3) The yeast cells have begun to utilize the CO_2 and water to carry out photosynthesis.
4) The alcohol produced has inhibited the growth of the yeast cells.
5) After 4 weeks the cells switched to aerobic respiration.

19. A student measured the amount of energy in a given amount of ADP and in an equivalent amount of ATP by two separate measurements with a bomb calorimeter. She then wanted to calculate the amount of energy transferred to one of these substances from glucose in respiration. Which of the following calculations would she perform?

ATP = energy per molecule of ATP
ADP = energy per molecule of ADP

1) ATP/ADP
2) ADP - ATP
3) ATP + ADP
4) ATP - ADP
5) ADP x ATP

20. Aerobic metabolism produces ATP much more efficiently than anaerobic metabolism. However, yeast cells grown under aerobic and anaerobic conditions with the same concentration of glucose in the medium, are found to contain the same concentration of ATP inside the cells. A biologist concludes that the cells use ATP at a slower rate under anaerobic conditions than under aerobic conditions. Which of the following assumptions is he most likely making?

1) Glucose metabolism is the only source of ATP in these cells.
2) The presence of oxygen stimulates the rate of ATP-consuming reactions.
3) Anaerobic metabolism produces less ATP from a molecule of glucose than does aerobic metabolism.
4) Anaerobic metabolism produced more ATP from a molecule of glucose than does aerobic metabolism.
5) These cells cannot carry out aerobic glucose breakdown.

21. Some general biology books say that aerobic respiration produces 36 molecules of ATP per molecule of glucose. Others say there are 38 ATP's per glucose. One says "aerobic respiration

yields anywhere from 21 to 36 ATPs per glucose molecule". Which of the following is the best response a student can make?

1) Conclude that the book which is most widely used is correct.
2) Wait until textbooks agree before studying respiration.
3) Conclude that all the authors are right depending on their assumptions.
4) Make tentative conclusions based on the whole range of possible numbers.
5) Ignore any estimates based on assumptions.

22. An isolated muscle contracts when ATP is added. It has also been observed that the amount of ATP in a muscle in an animal is approximately the same before and after a period of contractions. Which of the following hypotheses does not adequately interpret these observations?

1) ATP is the energy source for contraction and it is replaced as quickly as it is used.
2) Compound X is the energy source for contraction, and ATP prevents the release of energy from compound X.
3) ATP is not the normal energy source for contraction but removing the muscle damaged it so that ATP could be used.
4) Compound X is the energy source for contraction and the presence of ATP is required just to start the contractile processes, but ATP is not consumed.
5) None of the above are adequate explanations.

23. Student A hypothesized that plants obtained their energy from light and student B hypothesized that plants obtained their energy from nutrients in the soil. They set up the following experiment: Each of several groups of plants received different treatments of light and fertilizer. After all the seedlings grew for 21 days, the energy in each group of plants was measured using the bomb calorimeter. The treatments and results of the energy measurements are shown below:

Light Intensity	10	20	30	40	50
Fertilizer (grams)	1	2	3	4	5
Energy Content (KCal)	974	1190	1510	2170	2865

Which of the following would be the best statement about the data from the students' experiment?

1) The data contradict student A's hypothesis but support student B's hypothesis.
2) The data contradict both hypotheses.
3) The data contradict student B's hypothesis but support student A's hypothesis.
4) The data are consistent with both hypotheses.
5) The data show that energy is obtained from both light and fertilizer.

24. Student A of Question 23 did another experiment to test her hypothesis. She kept all variables constant and grew seedlings for 21 days. She carefully measured all the light energy striking the plants and the energy in the plants. She obtained the following results:

Energy in plants at start of experiment = 247 KCal.
Total energy in light during 21 day period = 24,743 KCal.
Energy in plants at end of 21 day period = 1743 KCal.

Which of the following would be the best statement student A could make?

1) The plants could be obtaining their energy from the light.
2) Energy in the plants is derived from the light absorbed but not from the fertilizer.
3) Light cannot account for the energy in the plants at the start of the experiment.
4) There is so much more energy in the light than in the plants that the plants must be obtaining energy from some other source.
5) 21 days is too short a period to obtain any meaningful evidence.

25. When comparing the changes in energy content and biomass in albino and green corn it is

frequently assumed that the only difference between the two types of corn is that albino corn lacks chlorophyll and cannot absorb light energy. Which of the following observations would provide the greatest support for the assumption?

1) The albino plants can be kept alive by applying a sugar solution to the leaves.
2) Albino plants appear very seldom in a natural population of corn and when they do appear, they quickly die.
3) When exposed to equal intensities of illumination the albino plants will die after about 10 days while normal corn will survive.
4) Since it has been shown that chlorophyll is essential for photosynthesis, the albino plant could not carry out photosynthesis.
5) The seeds produced using normal and albino plants are indistinguishable when visually inspected.

26. A visitor from another planet came to earth and observed humans. The visitor observed that humans were normally active during daylight hours and tended to become inactive at night. Each night as the dark period went on, more and more humans collapsed until daylight. The visitor hypothesized that humans derive their energy directly from sunlight falling on their bodies. Which of the following observations would also be consistent with the visitor's hypothesis?

1) Humans have decreased activity during a solar eclipse.
2) A pigment was isolated from human skin which absorbed some blue and red light.
3) Humans are less active on rainy days.
4) When the sun goes down, some humans switch on auxiliary light sources.
5) All of the above.

27. We can see transmitted or reflected light but not that which has been absorbed. A solution of chlorophyll appears green to our eyes. Which of the following conclusions can be reached from these observations?

1) Chlorophyll does not absorb much green light.
2) Chlorophyll does absorb light of some colors.
3) Chlorophyll absorbs green light.
4) 1 and 2 above.
5) 2 and 3 above.

28. A student wanted to measure an action spectrum for some corn seedlings. Which of the following would she do?

1) Measure the amount of chlorophyll present in leaves of the plants at different ages.
2) Extract the chlorophyll from the plants and measure the absorption of light of different wavelengths.
3) Measure the amount of light reflected from the leaf surface for different wavelengths.
4) Measure the increase in height of the plants when grown in light of different wavelengths.
5) Both 2 and 3 must be done.

29. A student was investigating the absorption of light by plants and the subsequent production of starch by the plants. Chemical analysis of the plant's leaves indicated the presence of two pigments: Pigment A which absorbed light of wavelength A and Pigment B which absorbed light of wavelength B. When light of wavelength A *only* was used to illuminate the plant, he observed a significant increase in the plant's starch content. He concluded that pigment B was not essential for photosynthesis. Which of the following would be the best criticism of his conclusion?

1) The data indicate that both pigments A and B may be necessary for photosynthesis.
2) Energy transfer to pigment B may be necessary before the energy can be used in photosynthesis.

3) Since pigment A and pigment B absorb different wavelengths of light, it is possible that only one would be necessary for photosynthesis in the plant.
4) Pigment B would not be present if it were not necessary for photosynthesis.
5) Both 3 and 4 are equally good criticisms of his conclusion.

30. The student in Question 29 later discovered two mutants of the plant used previously. Mutant A contained only pigment A and mutant B contained only pigment B. He then illuminated mutant A with light of wavelength A and mutant B with light of wavelength B. In neither plant did he observe a significant increase in the starch content. Which of the following would be the best conclusion?

1) Pigment B is not essential for photosynthesis.
2) Energy transfer from pigment A to pigment B is necessary before the energy can be used in photosynthesis.
3) Both pigments must absorb light which is used in photosynthesis.
4) Since pigment B was not present in mutant A it would not be expected to be capable of photosynthesis.
5) If mutant A had been illuminated with both wavelengths of light a significant starch increase would have been observed.

31. To test the hypothesis that something in the lower surface of leaves is responsible for the production of starch when exposed to light, the following experiment was performed. One leaf of a plant previously kept in the light was coated on top with vaseline as a control and one leaf was coated with vaseline on the bottom. The plant was then placed under lights for six hours. At the end of the six hours, both leaves were tested for starch. Both leaves gave a positive test with iodine. The best criticism of this experiment is:

1) The control plants were left under light so long that light eventually filtered through the bottom of the leaf.
2) The control should have been a leaf that had not received vaseline.
3) A plant previously kept in the dark should have been used at the start of the experiment.
4) More than one leaf should have been used for each treatment.
5) None of the above criticisms are valid and it was a good experiment.

32. When certain algae are exposed to radioactive carbon dioxide and light, starch is formed which contains radioactive carbon. These algae also form starch in the dark, when no carbon dioxide is present, if they are placed in solutions of glucose or glycerine (a three-carbon sugar). One biologist concluded from these observations that starch is not the first product of photosynthetic carbon dioxide uptake. How would you criticize his conclusion?

1) Photosynthesis cannot occur in the dark.
2) The plants might have converted the glucose or glycerine to carbon dioxide and then synthesized starch from it by photosynthesis.
3) Starch is insoluble. Since photosynthesis occurs in solution, carbon dioxide must first go into a soluble product, then into starch.
4) Starch might be formed by alternative routes: directly from carbon dioxide in light and from larger molecules in the dark.
5) Starch is too complex to be built up directly from carbon dioxide and water. Simple intermediates must exist.

33. When an illuminated green plant was exposed to radioactive carbon dioxide it was observed that radioactive carbon appeared in the starch produced by the plant. From these results a scientist proposed that the oxygen in carbon dioxide will also appear in the starch. Which of the following was he most likely assuming?

1) The oxygen and carbon in carbon dioxide are not separated in photosynthesis.

$$CO_2 + H_2O \rightarrow glucose + O_2$$

building blocks → respiration

2) The oxygen in the water added will not appear in the water produced by photosynthesis.
3) The hydrogen and oxygen in water cannot be separated.
4) The oxygen gas produced in photosynthesis is produced entirely by the plant.
5) Both oxygen gas and starch are produced by photosynthesis.

34. Many experiments on photosynthesis are interpreted as evidence that oxygen is produced in photosynthesis. Which of the following would provide the greatest support for this interpretation?

1) Greater oxygen production occurs in plants illuminated with red and blue light than with green light.
2) Water is a reactant as well as a product in the process of photosynthesis.
3) At least some of the oxygen in carbon dioxide is used in the formation of starch.
4) The leaf surface contains pores through which carbon dioxide and oxygen can move.
5) If plants are exposed to $H_2^{18}O$, labelled oxygen does not become incorporated into starch.

35. A student predicted that if part of a leaf is shaded and other parts of the leaf are fully illuminated, then equal amounts of starch will be found later in the shaded and illuminated parts. Which of the following is this student most likely assuming?

1) There is chlorophyll present in both the shaded and unshaded parts of the leaf.
2) In the shaded part of the leaf photosynthesis is increased.
3) Carbon dioxide and water can enter both the shaded and illuminated leaf cells.
4) By shading part of the leaf, photosynthesis is increased in illuminated parts of the leaf.
5) Starch can move to different parts of the leaf.

36. If a photosynthesizing plant is exposed to $H_2^{18}O$ in a sealed chamber for several days, respiration will also be occurring during this period. Thus, the products of photosynthesis will be used in respiration and the products of respiration will be re-used in photosynthesis, perhaps through several cycles. At the end of the exposure period, where would you expect to find ^{18}O?

1) H_2O
2) O_2
3) Both 1 and 2.
4) Neither 1 nor 2.
5) No prediction is possible.

37. A green plant is grown in normal air under constant illumination. Assuming the plant is respiring and photosynthesizing, which of the following predictions could you make?

1) Oxygen consumption by the plant will equal oxygen production.
2) Carbon dioxide consumption by the plant will equal carbon dioxide production.
3) The biomass of the plant will increase as respiration increases.
4) The change in biomass of the plant will equal the amount of oxygen consumed.
5) The biomass will change proportional to the amount of oxygen consumed or produced.

38. A plant physiologist grew some plants in a sealed chamber with alternating 12-hour light and 12-hour dark periods for several days. During this time she measured the amount of oxygen produced in the light periods and oxygen consumed in the dark periods as shown below:

Average oxygen increase in light period	12.5 units
Average oxygen decrease in dark period	8.1 units

She then concluded that photosynthesis produced 12.5 units of oxygen during a 12-hour light period. Which of the following is she most likely assuming?

1) Respiration does NOT occur in these plants in the light period.
2) There is a net production of oxygen during a 24 hour period.
3) Photosynthesis produces more oxygen than is consumed during a 24 hour period.
4) Respiration continues at the same rate in the light period and dark period.
5) In these plants photosynthesis can produce enough oxygen to meet the requirements of respiration.

39. A second plant physiologist examined the data in Question 38. She suggested applying an inhibitor of respiration to these plants which would not affect photosynthesis. She predicted that if such an inhibitor were applied, the amount of oxygen produced during the light period would be 20.6 units. Which of the following is she most likely assuming?

1) Respiration does not occur in the uninhibited plants in either the light period or dark period.
2) Respiration does not occur in the light period in uninhibited plants.
3) Respiration does not occur in the dark period in uninhibited plants.
4) Respiration continues at the same rate in the light and dark period in uninhibited plants.
5) Respiration increases during the light period in uninhibited plants.

40. Some evidence has indicated that green plants actually increase respiration when illuminated. If so, which of the following would most likely be the amount of oxygen produced by the plants in Question 38?

1) 24.8 units.
2) 20.6 units.
3) 8.1 units.
4) 4.4 units.
5) 12.5 units.

41. A green plant was placed in an atmosphere containing radioactive carbon dioxide and illuminated. Later the atmosphere was flushed and replaced with non-radioactive air. The light was continued. Suppose that this technique was used to test the hypothesis that respiration and photosynthesis occur simultaneously in the light. Which of the following sets of data would provide the strongest support for the hypothesis?

Amounts of radioactive CO_2 in atmosphere 0, 4 and 8 hours after flushing are shown below. Zero time measurement was taken immediately after flushing.

	0 hr.	4 hrs.	8 hrs.
1)	0	0	0
2)	14.6	13.9	13.1
3)	13.1	13.9	14.6
4)	0	18.3	36.4
5)	16.2	0.1	0.1

42. A plant physiologist was experimenting with plants which had previously incorporated radioactive carbon into starch stored in the plant cells. He measured the radioactivity present in the following compounds:

Sugar extracted from the plant
CO_2 produced by the plant
Alcohol, if produced by the plant

If he assumed that only aerobic respiration is occurring, where would he predict radioactivity would be found?

1) Sugar, CO_2 and alcohol.
2) Sugar and alcohol only.
3) Sugar and CO_2 only.

4) CO_2 and alcohol only.
5) CO_2 only.

43.   When the plant physiologist in Question 42 made his measurements, he found radioactivity in all three compounds. Which of the following would he most likely conclude?

1)   Anaerobic respiration is occurring.
2)   Only aerobic respiration is occurring.
3)   Only anaerobic respiration is occurring.
4)   Aerobic respiration could be occurring.
5)   Both 1 and 4.

44.   Then the plant physiologist in Question 42 placed the plants in the dark in an atmosphere containing no oxygen. Where would he most likely find radioactivity?

1)   Sugar, CO_2, and alcohol
2)   Sugar and alcohol only (Since there is no oxygen, no CO_2 can be produced.)
3)   Sugar and CO_2 only
4) CO_2 and alcohol only
5) CO_2 only

45.   A scientist placed a green plant in a sealed atmosphere of CO_2 containing radioactive carbon and watered it with labeled water ($H_2{}^{18}O$). After 12 hours in the light which of the following would he most likely find?

1)   Radioactive starch, no $^{18}O_2$
2)   Radioactive starch, $^{18}O_2$ and radioactive CO_2
3)   Radioactive CO_2, $^{18}O_2$ but no radioactive starch
4)   No labelled products in the atmosphere surrounding the plant
5)   A dead plant

46.   A sealed aquarium containing only plants, fish and bacteria was studied for two months. The biomass of each group of organisms stayed nearly constant during the study. Oxygen levels were higher during the day than at night, but the daily pattern of O_2 change was the same every day throughout the two months. The amount of O_2 produced by photosynthesis in 24 hours would be closest to which of the following?

1)   The O_2 consumed by bacteria and fish at night.
2)   The O_2 consumed by bacteria, fish and plants at night.
3)   The O_2 consumed by bacteria and fish in 24 hours plus that consumed by the plants at night.
4)   The O_2 consumed by bacteria, fish and plants in 24 hours.
5)   The O_2 consumed by bacteria and fish in 24 hours.

47.   For several weeks a student maintained a sealed aquarium containing some green plants and a single goldfish. If the student now added a second goldfish to the aquarium, which of the following would most likely occur?

1)   Oxygen concentration would reach a higher peak during the day.
2)   Oxygen concentration would decrease less during the night.
3)   Oxygen concentration would increase more slowly during the day.
4)   There would likely be no change in the oxygen concentrations.
5)   Oxygen concentration would decrease more slowly at night.

DIFFUSION AND TRANSPORT

Passive diffusion is the movement of material which occurs as a consequence of the constant random motion of molecules and atoms. Since random movement is increased when energy (heat) is added, the rate of diffusion of a substance also increases when the system is heated. Random motion also leads to the prediction that more molecules will move from regions of high concentration to regions of low concentration than will move back simultaneously. Movement across membranes will be affected by membrane permeability. The movement of water across membranes (osmosis), because it can result in relatively large cell volume changes, is considered a special case of diffusion, although it exhibits the same dependence on concentration and permeability. In many cases substances may be actively transported across membranes from a lower to a higher concentration. This requires the utilization of chemical energy (ATP) for transport.

Exchange of material, such as O_2 and CO_2, between a cell and the extracellular environment depends on the surface-volume ratio. In most multicellular organisms exchange is further enhanced by the bulk movement of extracellular fluid, usually in a circulatory system or in plants via a vascular system. Fluid in a circulatory system is moved by a pump or heart and passes through some organ of gas exchange, such as a lung or gill. Material is constantly being added to or removed from the system by gas exchange, digestive, and excretory organs and the contents of specific substances vary in predictable ways throughout the system. Pressure and volume flow also vary due to friction and the structure of the system.

In a kidney, urine is produced by filtration of the blood (a function of pressure and permeability), and by modification of the filtrate by reabsorption and secretion (functions of active transport, osmosis and diffusion).

EXAMPLE 1

A plant cell is exposed to an aqueous solution having sodium and chloride concentrations higher than the sodium and chloride concentrations of the cell interior. However, after 30 minutes a student observes the cell in a microscope and states that the cell shows no evidence of having lost water. Which of the following is the best explanation of his observation?

1) The cell membrane is not permeable to either sodium or chloride.
2) The cell has a rigid wall around it.
3) The cell protected itself by actively transporting sodium across its membrane into the external solution.
4) The plant cell, being different from an animal cell, cannot tolerate high concentrations of sodium and chloride and dies within 5 minutes.
5) The cell interior contains a great variety of solutes.

ANALYSIS:

1. If the cell membrane was permeable to sodium or chloride, these substances would tend to enter the cell by diffusion. This would tend to compensate for any volume changes caused by water leaving the cell. But if the membrane was not permeable to these substances, only water would move and shrinkage would be evident.

2. A rigid wall might prevent visible swelling but the cell could still shrink within the wall, leaving space between the cell membrane and the wall.

3. If the cell actively transported sodium out of itself, the concentration of water would become even greater inside and less outside the cell and the tendency to shrink from water loss would also be greater.

4. Whether the cell is dead or alive should not affect its permeability to water.

5. The tendency for water to leave the cell results from the higher concentration of water inside than outside. The concentration would be lower and the tendency for water to move out would be reduced. If the concentration of water inside was equal to that outside, the net movement of water would be zero.

EXAMPLE 2

A student hypothesized that the starch that appears in potato leaves during daylight is translocated to the leaves from potato tubers. Which of the following would provide the greatest support of his hypothesis?

1) When potato plants are exposed to radioactive CO_2 radioactive starch is later found in the tubers.
2) Starch is present in both the leaves and tubers of potato plants during daylight hours.
3) Potato tubers contain a greater concentration of starch than leaves.
4) Radioactive organic carbon compounds are found in the stem tissue when radioactive starch is present in both leaves and tubers.
5) In potato plants with radioactive starch originally present only in the tubers, some radioactive starch is later found in the leaves.

ANALYSIS:

1. If radioactive starch is found in the tubers, it was either formed there from radioactive CO_2 or

translocated there from some other site of synthesis. If the starch was formed in the tuber it might or might not be translocated to the leaves. This observation is consistent with both possibilities and thus is not evidence in support of either.

2. If starch is synthesized in the tuber and translocated to the leaves, it should be found in both places. But it should also be found in both places if it is synthesized in the leaves and translocated downward. This observation is also consistent with both possibilities.

3. If translocation was a passive process based on diffusion, then this concentration difference would contradict the hypothesis that starch was translocated from leaf to tuber and would thus give some support to the student's hypothesis. However, translocation is an active pumping of fluid. Furthermore, the starch is translocated as glucose monomers whose concentration can be regulated independently of the concentration of starch. Thus this concentration difference does not support either hypothesis.

4. Some radioactive compounds must be found in the stem if starch is being translocated between tubers and leaves. However, their presence does not, by itself, indicate the direction of translocation. Either hypothesis could be valid if this observation was made.

5. This is direct evidence that radioactive atoms moved from tuber to leaves. As such it is the only one of these observations which provides any evidence in support of the hypothesis. However, alternative interpretations still exist. The radioactive starch could have been broken down by the tuber cells, released into the atmosphere as radioactive CO_2 and incorporated into starch by the leaves.

1. Two sugar solutions of equal concentration are separated by a membrane which is permeable to sugar molecules. Some of the sugar molecules on the right side of the membrane contain radioactive carbon. What would you predict would be the final distribution of radioactive sugar on the two sides of the membrane?

   1) All radioactive sugar will be on the right side.
   2) More of the radioactive sugar will be on the right side than the left.
   3) The concentration of radioactive sugar will be equal on the two sides of the membrane.
   4) More of the radioactive sugar will be on the left side.
   5) All the radioactive sugar will be on the left side.

2. Using the same experimental arrangement as in Question 1, the sugar solution on the right side of the membrane was heated and the solution on the left side was cooled. Assuming there is no movement of water across the membrane, which of the following would most likely be the final distribution of sugar molecules on the two sides of the membrane?

   1) All sugar will be on the right side.
   2) There will be more sugar on the right side than on the left.
   3) There will be equal amounts of sugar on both sides.
   4) There will be more sugar on the left side than on the right.
   5) All the sugar will be on the left side.

3. Using the apparatus in Question 2, the solution on the left side was heated to the same temperature as the solution on the right side. Which of the following would most likely be the final distribution of sugar molecules?

   1) All the sugar will be on the right side.
   2) There will be more sugar on the right side than on the left side.
   3) There will be equal amounts of sugar on both sides.
   4) There will be more sugar on the left side than on the right side.
   5) All the sugar will be on the left side.

4. Two compartments were separated by a membrane with unknown permeability properties. To compartment A was added a 10% aqueous salt solution and to compartment B was added pure water. A student predicted that the solution in compartment A would be less than 10% salt if measured the next day. Which of the following was this student most likely assuming?

   1) The membrane is permeable to water only.
   2) The membrane is permeable to salt only.
   3) The membrane is permeable to both water and salt.
   4) The membrane is not permeable to salt or water.
   5) Choices 1, 2, and 3 are equally likely.

5. When some students reported to their lab they found an experimental apparatus consisting of two chambers separated by a membrane. Both chambers contained equal volumes of sugar solution but the solution in the left chamber contained twice as many sugar molecules as the right chamber. The students did not know when the apparatus had been set up. Which of the following would be most justified?

   1) Originally all sugar molecules were in the right chamber.
   2) Originally all sugar molecules were in the left chamber.
   3) Originally sugar molecules were equally distributed between the two chambers.
   4) The final distribution of sugar molecules will be equal between the two chambers.
   5) There is insufficient information to determine whether the membrane is permeable to sugar.

practice

6. A biologist prepared a strip of agar (a gelatin-like substance) as shown in the figure below. At position 4 she placed a concentrated sample of amino acid A and at position 9 a concentrated sample of enzyme E. E is known to catalyze the conversion of A to a product P. Assume that A and E can diffuse through agar and that smaller molecules can diffuse much more rapidly than larger molecules. At what position would you most likely first find P?

	Amino Acid A							Enzyme E	
1	2	3	4	5	6	7	8	9	10

1) 7
2) 6
3) 3
4) 4
5) 5

7. A student placed a dialysis bag containing a 10% glucose solution into a beaker also containing a 10% glucose solution. The dialysis bag was permeable to both glucose and water. If the student measured glucose movement out of the bag and glucose movement into the bag, which of the following would she most likely find?

Rate of Glucose Movement

	Out	In
1)	25	0
2)	25	25
3)	0	0
4)	50	0
5)	0	25

8. The student in Problem 7 now added to the solution in the beaker an enzyme which rapidly broke down glucose. Which of the following would most likely occur? Assume the enzyme did not affect the dialysis bag and did not appreciably change the water concentration.

1) Glucose movement into the bag will increase.
2) Glucose movement into the bag will decrease.
3) Net movement of glucose across the dialysis bag membrane will increase.
4) Both 1 and 3.
5) Both 2 and 3.

9. An apparatus consists of two chambers separated by a membrane. Assume that the membrane is permeable to water but not to solute. On the left side is pure water. On the right side is an equal volume of starch solution. Which of the following is the best prediction of the results several hours later?

1) There will be more water on the left than on the right.
2) There will be equal volumes of water on the two sides.
3) There will be more water on the right than on the left.
4) There will be equal amounts of carbohydrate on the two sides.
5) Both 2 and 4 are equally likely.

10. The same apparatus is set up in the same way with starch and water. In addition, the chamber on the right side of the membrane contains some water labelled with radioactive hydrogen atoms. Which of the following is the best prediction of the results hours later?

1) All of the radioactive water will be on the right.
2) Some of the radioactive water will be on the left.

3) Most of the radioactive water will be on the left.
4) All of the radioactive water will be on the left.
5) Both 1 and 4 are equally likely.

11. When a dialysis bag is filled with a dilute starch solution, securely tied, and placed in pure water the bag does NOT burst. Which of the following is most likely occurring?

1) Water will diffuse out of the bag but will not diffuse in.
2) Starch could make the dialysis bag more permeable to water movement in both directions.
3) Water will diffuse into the bag but will not diffuse out.
4) Increased pressure inside the bag causes more frequent collisions of water molecules with the dialysis bag membrane.
5) You would not expect the bag to burst based on differences in water concentrations.

12. Salt has been used for centuries to cure meat, fish and other foods. Assuming that food spoilage is caused by microorganisms, salt's effectiveness is a result of which of the following?

1) Salt draws water from the atmosphere causing the food to stay too moist to support bacteria.
2) The food is too salty to be attractive to microorganisms.
3) Salt is taken into the food by active transport thus raising the salt content to a high level.
4) Any microorganism coming in contact with the food will lose water and die.
5) The membranes of the cells of the meat become hard and impermeable.

13. Paramecium is a common one-celled animal in fresh water. The inside of a paramecium contains water with proteins, glucose, and various other molecules and ions dissolved in it. Assume the cell membrane of the paramecium is permeable to water and that the water surrounding the paramecium is pure. In a healthy living paramecium diffusion or random movement of molecules should be able to account for:

1) The water molecules moving into the cell, but not those leaving the cell.
2) The water molecules leaving the cell, but not those entering it.
3) All of the water molecules entering the cell, but only part of those leaving it.
4) All of the water molecules entering the cell and all of the water molecules leaving it.
5) The fact that the concentration of glucose inside the cell is greater than that outside the cell.

14. A student was experimenting with a cell type which could change its form to assume a variety of shapes without any volume change. When she maintained a culture of individual cells in a solution with a high oxygen concentration, they were shaped like discs about 1 unit thick and 5 units in diameter. If the cells were now transferred to a solution with a much lower oxygen concentration, which of the following shapes would you predict?

1) The discs would become much, much thinner.
2) The discs would become shaped like spheres or balls.
3) The cells would extend thin projections like fingers from their surface.
4) Both 1 and 2 could likely occur.
5) Both 1 and 3 could likely occur.

15. A physiologist was studying some cylindrical cells which were 2 mm in length and 0.02 mm in diameter. With these dimensions the cells apparently obtain sufficient oxygen. As these cells grow longer, which of the following will most likely occur?

1) Assuming the respiration rate increases, the cell can obtain sufficient oxygen.
2) The cell will experience significantly greater difficulty obtaining oxygen.

3) Since the larger cell will have a greater total respiration, it will require less oxygen.
4) As the cell becomes larger, the surface area will decrease.
5) The cell will experience no greater difficulty obtaining oxygen.

16. A cell biologist observed and measured the surface area and volume of a cell at various times, obtaining the data shown below:

Observation	1	2	3	4	5
Surface area	250	200	400	380	600
Volume	250	250	300	310	500

At which observation should the cell be able to most easily obtain sufficient oxygen to satisfy the requirements for respiration?

1) 1
2) 2
3) 3
4) 4
5) 5

17. In order to be able to decide on an answer for Question 16 which of the following assumptions must be made?

1) Permeability of the cell membrane to oxygen is the same at all observation times.
2) Respiration occurs at an equal rate throughout the cytoplasm.
3) The respiration rate of the cell is the same at all observation times.
4) 2 and 3 only must be made.
5) 1, 2 and 3 must be made.

18. Assume you have two cubes. One is 1 cm on a side, the other is 10 cm on a side, and you are concerned with the surface/volume relationships between these two cubes. Which of the following is the most valid statement?

1) The ratio of surface to volume in the 1 cm cube is 6 to 1.
2) The volume of the 10 cm cube is 1000 times greater than the 1 cm cube.
3) The 10 cm cube has a surface 100 times greater than the 1 cm cube.
4) The 1 cm cube, if it were an organism, would exchange more material with its environment per unit of volume than the 10 cm cube, because of its larger surface/volume ratio.
5) All of the above statements are correct.

19. A sponge is a saclike animal composed of layers of cells surrounding an inner water-filled cavity. Sponges feed on minute food particles which they capture in currents created by cells bearing whiplike structures called flagella. The amount of food a sponge can obtain depends upon the number of flagellated cells lining its internal surface. Large sponges need more food than smaller ones, and large sponges have thicker walls. How else should large sponges differ from smaller ones?

1) Large sponges are more spherical in shape than smaller ones.
2) Large sponges have numerous folds in the inner surface.
3) As the sponge becomes larger its total surface area decreases.
4) As the sponge becomes larger its volume does not increase as much as its surface.
5) Assuming the sponge is approximately cube shaped, doubling its volume will also double its area thus helping to maintain an optimal surface/volume ratio.

20. Certain kinds of salt water algae have iodine concentrations in their cells that are many times

higher than the iodine concentration in sea water. Which of the following can account for this?

    1)   Iodine is accumulated by the algal cells by active transport.
    2)   Iodine is incorporated within the cells into polymerized compounds.
    3)   Iodine can only cross the cell membrane with carbon dioxide. When the CO_2 is used there is nothing to transport iodine out again.
    4)   1 and 2 are equally effective.
    5)   All of the above are equally effective.

21.   A student wanted to determine the permeability of yeast cell membranes to phosphate ions. He added a known amount of phosphate ions to a solution containing yeast cells, waited 24 hours, and then measured the concentration of phosphate ions remaining in the solution. He found that the concentration of phosphate ions in the solution had not changed in the 24 hours. The student concluded that the yeast cell membranes were not permeable to phosphate ions. Which of the following is the best criticism of the student's conclusion?

    1)   Phosphate ions must cross the cell membrane because phosphate ions are found inside the cell.
    2)   Equal amounts of phosphate ions may be diffusing in and out of the cell so that there is no apparent change in the phosphate ion concentration of the external solution.
    3)   Phosphate ions may be diffusing into the cells and then may be incorporated into larger organic molecules.
    4)   Phosphate ions are being actively pumped out of the cells.
    5)   A permeable cell membrane will allow water and solutes to diffuse in and out of the cell.

22.   In reference to Question 21, which of the following would be the best test of yeast cell membrane permeability to phosphate?

    1)   Add radioactive phosphorus to the external solution, wait 24 hours, then measure the radioactivity inside the cells.
    2)   Add radioactive phosphorus to the external solution, wait 24 hours, then measure the concentration of phosphorus in the solution.
    3)   Add a chemical which inhibits ATP in the cell, wait 24 hours, then test the cells for the presence of phosphate ions.
    4)   Add phosphate ions plus a chemical which inhibits ATP to the solution, wait 24 hours and measure the concentration of phosphate ions in the cells.
    5)   Add phosphate ions plus a chemical which inhibits ATP to the solution, wait 24 hours and then measure the concentration of phosphate ions in the external solution.

23.   When given plenty of glucose, some cells may store the surplus as glycogen, which is a polymer made up of glucose molecules. The glycogen may be broken down and used later if the cells run out of glucose. The advantage to the cells of converting extra glucose to glycogen is most likely that:

    1)   Lowering the level of free glucose inside the cell would facilitate the net movement of more glucose into the cell by diffusion.
    2)   Glycogen molecules, being much larger than glucose, are less likely to be lost by diffusion across the membrane.
    3)   It is easier for cells to keep inventory on stored glucose molecules.
    4)   Both 1 and 2 are advantageous.
    5)   1, 2 and 3 are advantageous.

24.   Most cells can maintain a low internal sodium ion concentration even though the external sodium ion concentration is high. This observation has led many scientists to conclude that sodium ion is actively transported out of the cell. Which of the following were they assuming?

1) The cell membrane is permeable to sodium ion.
2) Sodium ion is actively transported into the cell.
3) Sodium ion is "bound" inside the cell.
4) Sodium ion is actively transported out of the cell.
5) Either 2 or 3 is probably being assumed.

25. A researcher added red dye to a mixture of dead and live cells. About half of the cells remained clear (white) while about half became red. Which of the following is the best explanation for this situation.

1) The cell membrane of the dead cells could not transport the dye, therefore the dead cells remained clear (white).
2) The live cells, using active transport, selectively moved dye molecules inside the cells, turning them red.
3) The cell membrane of the dead cells is dead also, and diffusion of the dye could occur, so that the dead cells turned red.
4) The live cells, using energy (ATP), could prevent the entry of the dye, therefore the live cells remained clear.
5) From these data all of the above are possible.

26. In nerve cells, as in almost all cells, sodium ion (Na^+) concentration is much lower than in the surrounding medium. When a substance known to inhibit ATP production is added to isolated nerve cells at time = 0, the following results are obtained:

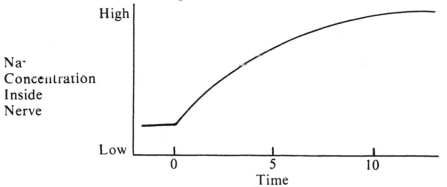

Na^+ in the surrounding medium is maintained at a high level. Which of the following are necessary to account for (i.e., "explain") these results?

A. The ATP inhibition causes the nerve cells membrane to become impermeable to Na^+.
B. Na^+ is "actively transported" out of the nerve cell.
C. The nerve cell membrane is permeable to Na^+.

1) A only.
2) B only.
3) C only.
4) A and B only.
5) B and C only.

27. It was found that a certain microorganism could utilize the amino acid histidine when it was supplied in the growth medium. A scientist proposes that the transport of histidine into the cells is catalyzed by an enzyme complex in the cell membrane. The initial rate of uptake of histidine (amount per cell per minute) can be readily determined after histidine is added to the growth medium. Which of the following observations would support the scientist's hypothesis?

1) The initial rate of uptake increases steadily as the concentration of histidine in the

medium increases.

2) At histidine concentrations above a certain value the initial rate of transport does not increase with additional increases in histidine concentration.

3) The initial rate of uptake of histidine is strongly dependent upon pH of the medium.

4) The initial rate of uptake decreases with higher concentrations of histidine.

5) Both 2 and 3 support the hypothesis.

28. Red blood cells maintain a high internal potassium ion (K^+) concentration and a low internal sodium ion (Na^+) concentration when in a bathing solution of high Na^+ and low K^+.

Bathing Solution		Red Blood Cells
K^+	5	95
Na^+	145	10

A student (A) hypothesized that these ion differences were maintained by a low membrane permeability and that lowering the temperature of the bathing solution and cells would increase permeability. If the student's hypotheses are valid, which of the following would most likely occur when the temperature is decreased?

1) Internal K^+ would increase.
2) Internal Na^+ would increase and internal K^+ would decrease.
3) Both internal K^+ and Na^+ would increase.
4) Internal Na^+ would decrease.
5) Both 1 and 4.

29. A second student (B) examined the information in Problem 28 and hypothesized that the internal ion concentrations were maintained by active transport of Na^+ and K^+ at the cell membrane. She predicted that lowering the temperature of the solution and cells would inhibit any active transport. Which of the following would this student predict when the temperature is lowered?

1) Internal K^+ would increase.
2) Internal Na^+ would increase and internal K^+ would decrease.
3) Both internal K^+ and Na^+ would increase.
4) Internal Na^+ would decrease.
5) Both 1 and 4.

30. Red blood cells lost K^+ and became "loaded" with Na^+ when placed in the appropriate bathing solution at lower temperature ($4°$ C). When so treated, the internal ion concentrations were: $K^+ = 5$, $Na^+ = 90$. These cells were then divided into two samples. Each sample was placed in a different bathing solution (A or B), and warmed to body temperature. The bathing solutions differed only in the concentrations of Na^+ and K^+. After 3 hours, the ion concentrations shown below were observed.

		Ion Concentrations	
Bathing Solution	Ion	In Bathing Solution	In Red Blood Cells
A	K^+	5	85
	Na^+	100	10
B	K^+	5	5
	Na^+	0	90

Which of the following would be the best interpretation of these data?

1) Warming increases the red blood cell membrane permeability to both ions.
2) Both Na^+ and K^+ are actively transported.

3) Active transport of Na^+ in one direction requires transport of K^+ in the other.
4) The red blood cell membrane is not permeable to Na^+.
5) Both 2 and 3.

The following information applies to Questions 31-35.

In salmon the principal permeable membrances exposed to the external water are the gills. Gill tissues are readily permeable to water and most ions. Young salmon hatch in fresh water and remain there for one or more years. After this initial period of growth they migrate to a seawater environment.

	Salmon Blood	Freshwater	Seawater
Sodium (Na^+)	140	1	450
Chloride (Cl^-)	3	1	20
Potassium (K^+)	2	1	10

31. Assuming that permeability is the only consideration, and active transport does not occur, which of the following best describes ion exchange between salmon blood and freshwater?

1) Blood gain of $Na^+ >$ gain of $Cl^- >$ gain of K^+.
2) Blood gain of $K^+ >$ gain of $Na^+ >$ gain of Cl^-.
3) Blood gain of $Cl^- >$ gain of $Na^+ >$ gain of K^+.
4) Blood loss of $Na^+ >$ loss of $Cl^- >$ loss of K^+.
5) Blood loss of $K^+ >$ loss of $Cl^- >$ loss of Na^+.

32. Assuming that permeability is the only consideration, and active transport does not occur, which of the following best describes ion exchange between salmon blood and seawater?

1) Blood gain of $Na^+ >$ gain of $Cl^- >$ gain of K^+.
2) Blood gain of $K^+ >$ gain of $Na^+ >$ gain of Cl^-.
3) Blood gain of $Cl^- >$ gain of $Na^+ >$ gain of K^+.
4) Blood loss of $Na^+ >$ loss of $Cl^- >$ loss of K^+.
5) Blood loss of $K^+ >$ loss of $Cl^- >$ loss of Na^+.

33. Assuming that permeability is the only consideration, and active transport does not occur, which of the following best describes osmosis between salmon blood and seawater.

1) Blood gain of water.
2) Blood loss of water.
3) Blood gain of water and salt except Na^+.
4) Blood loss of water and salts except Na^+.
5) Blood gain of water and salts except K^+.

34. Specialized cells develop in the gill tissues of salmon prior to their migration to the sea. These cells have an abundance of an enzyme which runs a "sodium pump" system. Which of the following is the most likely function of these special cells in salmon in seawater?

1) Passive transport (by diffusion) of Na^+ from seawater to the blood.
2) Passive transport (by diffusion) of Na^+ from the blood to seawater.
3) Active transport (requiring energy) of Na^+ from seawater to the blood.
4) Active transport (requiring energy) of Na^+ from the blood to seawater.
5) Active transport (by diffusion) of Na^+ from seawater to the blood.

35. Which of the following would you expect to be the most likely means of osmotic and ionic regulation? Note: Salmon kidneys cannot excrete urine more salty than the blood.

1) Salmon in seawater drink water, excrete excess salt from the gills and reduce excretion of water from the kidneys.
2) Salmon in seawater do not drink water, do excrete excess salt from the gills and do reduce excretion of water from the kidneys.
3) Salmon in freshwater drink water, excrete excess salt from the gills and increase excretion of water from the kidneys.
4) Salmon in freshwater do not drink water, do not excrete salt from the gills and do reduce excretion of water from the kidneys.
5) Both 2 and 3.

36. Some plants when grown in dry conditions exhibit characteristics different from those exhibited when the plants are grown with plenty of water. Which of the following would you expect to find in plants grown under dry conditions:

   A. Thin flat leaves with many pores.
   B. Small thick leaves with few pores.
   C. Large spreading root system.
   D. Root system confined to a very small area.
   E. Leaf pores which remain open constantly.
   F. Leaf pores which open at night.

1) A, C, and E
2) B, D, and E
3) B, C, and F
4) A, C, and F
5) A, D, and F

37. A physiologist measured the average movement of CO_2 from the blood into the lung as 125 units/minute. Average CO_2 movement from the lung to the blood was 35 units/minute. What was the net CO_2 movement?

1) 160 units/minute.
2) 90 units/minute.
3) 125/36 = 3.57 units/minute.
4) 125 x 35 = 4375 units/minute.
5) Cannot be determined from these data.

38. Consider these data from human lungs:

	Inhaled Air	Air in Lung	Blood
O_2	159 mmHg	104	70
CO_2	0.15	40	43

Which of the following is the best interpretation of these data?

1) Oxygen will diffuse out of the blood into the inhaled air; carbon dioxide will diffuse into the blood stream.
2) The pH of the blood will be too low for the person to survive because of the high CO_2 concentration.
3) Oxygen is used by the body's cells to help produce ATP in aerobic respiration.
4) Carbon is actively being pumped out of the blood and oxygen is being actively taken up by the blood.
5) Oxygen will diffuse from the inhaled air into the blood; carbon dioxide will diffuse from the blood to the inhaled air.

39. A student examined the following data on carbon dioxide concentrations in the external air, lung, and the blood:

	Air	Lung	Blood
CO_2	35	28	47

Why would the student most likely suspect that these data are in error?

1) The blood CO_2 should be lower than the lung CO_2.
2) The lung CO_2 should be higher than the air CO_2.
3) The blood CO_2 should be lower than the air CO_2.
4) The air CO_2 should be higher than the lung CO_2.
5) The CO_2 concentration in all three locations should be equal.

40. The best evidence to support the hypothesis that the lungs function as a gas exchange organ would be obtained by:

1) Measuring the CO_2 concentration difference between blood entering and leaving the lungs.
2) Measuring the CO_2 concentration difference between inhaled and exhaled air.
3) Measuring the surface area of the air sacs.
4) Both 1 and 2.
5) Measuring O_2 consumption by lung tissue.

41. Which of the following would increase the net movement of O_2 from the lung into the blood?

1) Increasing O_2 concentration in air drawn into the lung.
2) Decreasing the permeability of the lung to O_2.
3) Absorbing O_2 in a non-diffusible form once it enters the blood.
4) Both 1 and 3.
5) Both 2 and 3.

42. The rate of transpiration in a plant growing under constant conditions in a sealed container is determined. Which of the following changes would produce the greatest increase in the rate of transpiration over a long period of time?

1) Adding CO_2 to the atmosphere of the plant.
2) Adding minerals to the soil of the plant.
3) Decreasing the temperature in the plant's chamber.
4) Saturating the soil with water.
5) Removing humidity from the air in the chamber.

43. A biologist wondered about enzymes that break down proteins inside the human digestive system. "Why", he asked, "do they not break down proteins inside the cells where they are first synthesized?" Which of the following is the best answer to his question?

1) They do and that's why the liver is so big.
2) They don't because they are lost in the feces.
3) They do, thereby reducing the requirement for protein in the diet.
4) Maybe they do and maybe they don't, but if they didn't, we couldn't eat meat.
5) They don't because they are only active in the environment found inside the digestive tract.

44. Tadpoles, which are herbivores, have a much longer intestine than the adult frog which is a carnivore. Which of the following would you most likely predict from this observation?

1) Tadpoles have a lower energy requirement per unit body weight than do frogs.

2) Enzymes digesting plant material work more slowly than enzymes digesting animal material.
3) The intestine wall of the tadpole is much more permeable to digested food material than the intestine wall of the frog.
4) A frog requires a greater amount of food per unit body weight than a tadpole.
5) Tadpoles are smaller than frogs.

45. Termites live exclusively by eating wood which is made up primarily of cellulose. Certain one-celled organisms live only inside the gut of termites. When the one-celled organisms are removed from the termite gut, the termites soon die unless the one-celled organisms are replaced. Which of the following statements based on these observations would you be most likely to reject?

1) The one-celled organisms cannot absorb cellulose.
2) The termites cannot absorb cellulose.
3) The one-celled organisms cannot break down cellulose to glucose units.
4) The termites cannot break down cellulose to glucose units.
5) Both the termites and the one-celled organisms can absorb glucose.

46. Some cells are observed to engulf particles of material from their environment. The engulfed particles are then found to be enclosed in bags or vacuoles with a membrane around them. It is concluded that this process represents a feeding mechanism. Which of the following assumptions is not required in order to make the conclusion?

1) Digestive enzymes are released into the vacuoles.
2) All the contents of the vacuoles are completely digested, leaving no residue.
3) The monomers formed by digestion are absorbed across the vacuole's membrane into the cell.
4) The digestive enzymes cannot cross the vacuole's membrane into the cell.
5) The digestive enzymes cannot digest the vacuole's membrane.

Use these diagrams in Questions 47 through 50.

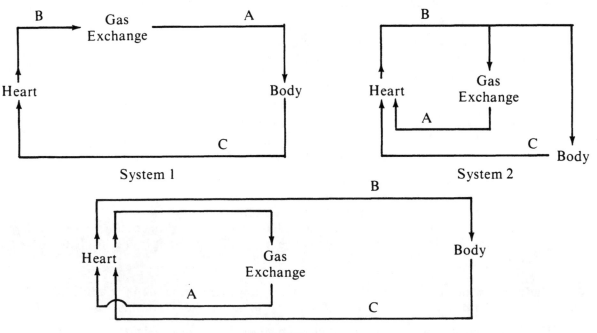

System 1

System 2

System 3

47. Which of the following would be the most likely statement regarding the oxygen concentration in blood at the points indicated? Assume equal rates of blood flow at point B, equal rates of gas exchange, and equal total rates of respiration in the three systems.

1) The concentrations are the same at point B in all 3 systems.
2) The concentrations are the same at points B and C in system 3.
3) The concentrations are the same at point A in systems 1 and 3.
4) The concentrations are the same at point B in systems 1 and 2.
·5) The concentrations at point C are the same in systems 2 and 3.

48. Which of the following would be the most likely statement regarding the volume of blood flow at points indicated? Use the same assumptions as in Question 47.

1) Flows are equal at A, B, and C in system 3.
2) Flows are equal at B and C in system 2.
3) Flows are equal at A in all three systems.
4) Flows are equal at A and C in system 2.
5) Flows are equal at C in all three systems.

49. Suppose the barrier through the middle of the heart in system 3 were removed. Which of the following statements would be most likely?

1) System 3 could obtain more oxygen from the atmosphere.
2) Systems 2 and 3 would now be very similar.
3) Blood flow at B would now be equal to flow at A in system 3.
4) Systems 1 and 3 would now be very similar.
5) Both 1 and 2 are equally likely as the best choices.

50. If the heart in system 2 is pumping 1.0 liter of blood each hour, how much is it pumping to the gas exchange organ?

1) 1.0 liter.
2) Less than 1.0 liter.
3) 0.5 liter.
4) 2.0 liters.
5) At least 0.5 liter, because the lungs require a great amount of O_2 in order to allow gas exchange to take place.

51. In the circulatory system of mammals blood enters the heart at 2 points and leaves the heart at 2 points. Which of the following would be the most likely measurements of oxygen concentration in the blood at the 2 entrances and 2 exits?

	Entrance 1	Entrance 2	Exit 1	Exit 2
1)	20	20	20	20
2)	100	100	20	20
3)	100	20	20	20
4)	20	100	20	100
5)	100	100	100	100

52. Blood is at 50 mm pressure each time it leaves the heart of a fish and is at 50 mm pressure each time it leaves the heart of a frog. In both the fish and the frog the pressure drops 20 mm each time it passes through a network of capillaries. For simplicity assume the only reduction in blood pressure is that which occurs when the blood passes through capillaries. The pressure in the veins which return blood to the heart of the fish should be:

1) Considerably less than in the frog.
2) The same as in the frog.

97

3) Greater than in the frog.
4) Approximately 50 mm.
5) Approximately 30 mm.

53. A scientist observed that blood leaving the heart contains the same concentration of oxygen as it did when it entered the heart. He proposed that the heart obtains its energy solely by anaerobic breakdown of glucose. Which of the following observations provides the best criticism of his proposal?

1) There is more CO_2 in blood leaving the heart than in blood entering.
2) Blood is brought to the cells of the heart by vessels which loop back after leaving the heart.
3) Anaerobic glucose breakdown by the heart would cause constant drunkenness.
4) There is less glucose in blood leaving the heart than in blood entering.
5) If the blood glucose level falls too low, the heart stops.

54. A student hypothesized that an efficient circulatory system would include cells in the blood that transport O_2 and CO_2. Which of the following characteristics would be most beneficial for the transporting cells to possess?

1) The cells release O_2 at high concentrations of CO_2.
2) The cells release CO_2 at high concentrations of O_2.
3) The cells take up CO_2 at low concentrations of CO_2.
4) The cells take up O_2 at low concentrations of O_2.
5) 1 and 2 are most beneficial.

55. A green plant that produces tubers is grown in total darkness for a period of 3 days. It is then moved into the light and supplied with radioactive carbon dioxide. At the end of 5 hours in the light which of the following would most likely be true.

1) There would be no detectable starch in the stem or tuber.
2) Radioactivity would be detected in leaf, stem and tuber.
3) Radioactivity would be detected in the leaf and stem but not in the tuber.
4) Starch would be detected in leaf, stem, and tuber.
5) There would be radioactive starch in the leaf and stem but not in the tuber.

56. Carnivores take in more salt in their diet than they require each day. They excrete the excess salt in the urine. Which of the following hypotheses would you make about a carnivore which produced urine containing a lower concentration of salt than the blood?

1) It lives in the desert and consumes very little water.
2) It is a sea water fish.
3) It has an abundant supply of fresh water and drinks freely.
4) Its kidneys do not reabsorb salt from the filtrate.
5) It loses a great deal of water from its gas exchange organs.

57. Certain dyes can be injected into the bloodstream and will subsequently appear in the urine. A thin slice of kidney can be incubated in an appropriate medium. If the dyes are added to the medium, they are observed to accumulate in the slices at higher concentration than in the medium. Kidney physiologists take these observations as evidence that the kidneys actively secrete these dyes into the urine at higher concentrations than occur in the blood. Which of the following assumptions are they making?

1) The dye is chemically changed by the cells in the slice.
2) The dye is accumulated within the slices in the spaces normally occupied by urine in the kidney.
3) The dye is present in the filtrate formed in the kidney.

4) The dye is not filtered in the slice.
5) Cell respiration does not occur in the slice.

The following information pertains to Questions 58 and 59.

The kidney functions by filtering the blood plasma and subsequently reabsorbing some solutes from the filtrate and secreting other solutes into the filtrate. The data in the table refer to the human kidney. Inulin and PAH are molecules which are soluble in water, easy to measure, and which have no physiological effects on the body.

Plasma Filtration Rate = 100 ml/min
Urine Flow Rate   =   5 ml/min

Solute	Concentration in Plasma	Concentration in Urine
Glucose	5 mg/ml	40 mg/ml
Inulin	5 mg/ml	100 mg/ml
PAH	0.1 mg/ml	12 mg/ml

Remember: Amount = Concentration x Volume

58. Which of the following mechanisms does the kidney use in processing glucose?

1) Filtration
2) Reabsorption
3) Secretion
4) Filtration and reabsorption
5) Filtration and secretion

59. Which of the following mechanisms does the kidney use in processing inulin?

1) Filtration
2) Reabsorption
3) Secretion
4) Filtration and reabsorption
5) Filtration and secretion

60. If a chemical which inhibits ATP was added to a mammalian kidney, which of the following would you expect to observe?

1) Increased reabsorption of solutes from the filtrate into the blood.
2) Decreased concentration of glucose in the urine.
3) Decreased amount of filtrate present.
4) Increased concentration of protein in the urine.
5) Decreased reabsorption of solutes from the filtrate into the blood.

61. The kidneys of a normal human produce about 180 liters of filtrate each day. This filtrate contains 552 grams of sodium, yet the daily intake from the diet and other sources is only 6 grams. A student says that this is ridiculously wasteful. He says that the kidney should only make enough filtrate to contain the 6 grams of sodium and save the extra energy of filtration and reabsorption. How would you criticize the student's ideas?

1) The body can't produce 180 liters of filtrate because it only contains about 3 liters of blood.
2) Glucose is completely reabsorbed from the filtrate.
3) About 1.5 liters of urine are formed from 180 liters of filtrate.
4) The concentrations of other substances in addition to sodium must be controlled in the blood.

5) Reabsorption of sodium and water takes place by passive diffusion and does not require energy.

62. The concentrations of most materials are equal in plasma entering the kidney and in the filtrate produced during the initial phase of blood processing. Protein, however, is found in the plasma but not in the filtrate. Therefore, the total concentration of all solutes (including protein) is higher in the plasma than in the filtrate. A student commented that a net flow of fluid from the blood to the filtrate would be impossible because water would tend to diffuse from the filtrate back into the blood. The student is mistaken because he failed to take into account which of the following?

1) Solutes may diffuse through water, but water itself does not diffuse.
2) Protein dissolved in water does not contribute to the total solute concentration.
3) The temperature of the filtrate is several degrees higher than that of the plasma entering the kidney, because the kidney maintains a very high rate of metabolism.
4) The pressure of the plasma entering the kidney is higher than the pressure of the filtrate.
5) The data given are incomplete. There must be some other substance, such as amino acids, present in the filtrate to bring its solute concentration up to that of plasma.

63. It has been observed that certain desert rodents eat a diet composed entirely of dry seeds, yet never drink water. Their kidneys produce a very concentrated urine, in order to eliminate waste with as little water as possible, yet more water is excreted than is contained in the dry seeds they eat. Which of the following is the best explanation of these observations?

1) Their kidneys do not function by the usual filtration and reabsorption mechanism.
2) They reproduce at a very young age because dehydration severely limits their lifespan.
3) The water deficit is replaced by water formed by cell respiration.
4) They produce very little metabolic waste.
5) We are assuming that no water is lost by evaporation from the rodent's body.

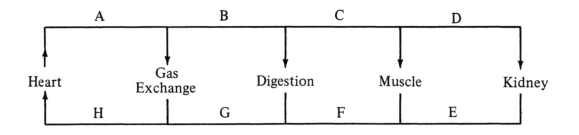

Which of the following sets of measurements would be most consistent with the hypothetical circulatory system shown above? Assume pressure, O_2 concentration and blood glucose level can change only at organs.

64. Blood pressure (mm Hg) at points indicated by:

	A	B	C	D
1)	26	18	16	12
2)	26	29	24	29
3)	26	26	26	26
4)	26	18	26	18
5)	26	2	2	2

65. O_2 concentration (ml O_2/liter) at points indicated by:

	A	C	F	H
1)	20	22	22	20
2)	20	22	11	19
3)	20	20	11	21
4)	20	22	11	11
5)	20	20	20	20

66. Flow rate (liters/hour) at points indicated by:

	A	D	E	H
1)	5	5	2	2
2)	5	2	5	2
3)	2	5	2	5
4)	5	2	2	5
5)	2	2	5	5

67. Blood glucose level (units) at points indicated by:

	B	C	F	G
1)	6	6	6	6
2)	6	8	8	6
3)	6	1	2	5
4)	6	3	3	6
5)	6	6	1	7

BIOLOGICAL INFORMATION

Reception and transfer of information from the external environment are essential for an organism's adaptive responses to the physical environment and other organisms. Light, sound, odor, and temperature are some of the stimuli which may be perceived by the sense organs or receptors of an organism. An organism's responses to controlled variations of environmental variables can lead to hypotheses regarding the internal mechanisms of the responses.

Within the organism the transfer and integration of information determine the responses. Many responses act to maintain homeostasis in variables such as blood pressure and flow, limb position, and blood glucose. In such cases the control mechanisms involve negative feedback. Most responses exhibit a threshold; if a stimulus is below the threshold, no response is elicited.

Transfer and integration within the organism occur via nerve cells, circulatory hormones, or a combination of the two. Hormones may be produced by specific organs in response to appropriate stimuli, and information is transferred by the type and concentration of hormones produced. The response also depends on the distribution of specific hormone receptors. More rapid and localized transfer and response are achieved by nerve cells or neurons. Information is transmitted by rapid electrical changes (action potentials) which rapidly move along the axon membrane.

The action potential is an all-or-none response but information can be coded by the frequency of action potentials. Communication occurs between nerve cells at synapses. At synapses, nerve cells do not touch but are separated by a gap, the synaptic cleft. Information transfer is accomplished by a transmitter substance which is released by one nerve cell and diffuses across the cleft to the other cell. In any neuron many incoming signals can be integrated into one outgoing signal. Both excitatory and inhibitory signals may be received and integrated.

Muscle tissue responds to stimuli from nerve cells. A single muscle cell or fiber exhibits an all-or-none response. Whole muscle, however, exhibits a graded response proportional to the stimulus strength.

EXAMPLE 1.

A biologist hypothesized that a bird's testes can sense day-length and then increase in size and produce a specific hormone called FSH. Increased levels of FSH in the bird's blood then lead to increased aggressiveness of the bird toward other male birds. Which of the following observations would represent the greatest contradiction of the hypothesis?

    1)   Testis size increases only when FSH is injected.
    2)   Birds become aggressive when only FSH is injected.
    3)   Testis size increases only when day length becomes longer.
    4)   Birds with testes removed do not become aggressive.
    5)   Both 2 and 4 are equally valid as the greatest contradiction.

ANALYSIS:

    A contradiction would be evident if a response was elicited which would not be predicted from the flow of information hypothesized above.

1. This response would not be predicted since no effect of FSH on testis size is hypothesized. According to the hypothesis, FSH is produced after the testes increase in size. Therefore, this is a contradiction and is a good choice.

2. This response would be predicted since increasing FSH is hypothesized to increase aggressiveness.

3. This response is consistent with the hypothesis. Day length affects testis size, according to the hypothesis.

4. Again, this response would be predicted. If testes are removed, no FSH should be produced and no aggressiveness should result.

5. Both 2 and 4 have been rejected.

EXAMPLE 2.

A certain person was not able to feel pressure on his hand. A biologist knew that there were two nerve cells between the pressure receptors and the brain, arranged like this:

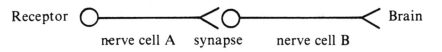

The biologist thought the trouble was in the synapse and made two hypotheses:

    A.   No transmitter is released from nerve cell A.
    B.   Nerve cell B does not have receptors for the transmitter.

Which of the following experiments would be most likely to allow him to decide between these two hypotheses?

    1)   Stimulate nerve cell A electrically and record from nerve cell B.
    2)   Apply chemical transmitter to the body of nerve cell A and record from nerve cell B.
    3)   Stimulate nerve cell B electrically and ask the person what he feels.
    4)   Apply chemical transmitter to nerve cell B at the synapse and record from nerve cell B.
    5)   Press on the receptor and record from nerve A.

ANALYSIS:

In order to decide between the two hypotheses, he needs a test for which each hypothesis predicts a different outcome.

1.   This will generate an action potential in cell A but both hypotheses predict that no response will be recorded in cell B.

2.   Cell A may or may not respond to its own chemical transmitter but even if it does, both hypotheses again predict no response from cell B.

3.   The hypotheses concern the transmission of information from cell A to cell B. Neither of them makes any prediction about the response of the brain when cell B is stimulated.

4.   If transmission is blocked by failure of cell A to release transmitter, then cell B should respond when the transmitter is added by the experimenter. On the other hand, if cell B does not have receptors, no amount of exogenous transmitter will elicit a response in cell B. For this test the two hypotheses predict different outcomes.

5.   Neither hypothesis makes a prediction about the mechanism of stimulation of cell A. They concern the transmission between cells A and B.

1. A student was studying an isopod and observed that it walked faster and turned more frequently in random directions as light intensity was increased. Which of the following would you predict if isopods were placed in a cage with light and dark areas?

   1) More isopods would be found in the light.
   2) Isopods would be equally distributed in light and dark areas since movement is random.
   3) More isopods would be found in the dark.
   4) Isopods would be equally distributed, but would spend more time in the dark areas.
   5) Isopods would stop walking when they reached the area between the dark and the light.

2. Three animals received the same signal from a fourth animal, the sender. Animal A approached the sender, animal B moved away from it, and animal C made no response. Which of the following statements is the best interpretation?

   1) Either A or B misinterpreted the intended message of the signal.
   2) The signal's message had separate meanings for A and B.
   3) The sender communicated with all three receivers.
   4) The sender did not communicate with C.
   5) The sender communicated with A and B.

3. Potato tubers are sometimes infected with a fungus which produces a chemical called an elicitor. In certain potatoes this elicitor causes the tuber cells to produce another chemical known as a phytoalexin. Potatoes which do not produce the phytoalexin are more rapidly and more completely infected with the fungus than the potatoes which do produce the phytoalexin. Which of the following could be a valid hypothesis regarding the resistance of potatoes which do produce the phytoalexin?

   1) The phytoalexin stimulates the activity of the elicitor.
   2) The phytoalexin inhibits the growth of the fungal cells.
   3) The phytoalexin inhibits aerobic respiration.
   4) The phytoalexin inhibits the growth of the potato tuber cells.
   5) The phytoalexin stimulates the growth of potato tuber cells.

4. Suppose a student discovered that coleoptiles bend toward a light source at the same rate regardless of the light intensity (i.e., bending rate is independent of light intensity). He used four light sources in his experiment, a 40 watt bulb, a 60 watt bulb, a 100 watt bulb, and a 200 watt bulb. Which of the following is the most likely statement regarding the information carried by light from each of the four light sources to the coleoptile?

   1) He cannot draw his conclusion using only four light sources.
   2) Light from the 200 watt bulb carries twice as much information as light from the 100 watt bulb.
   3) Light from a 100 watt bulb carries more information than a 60 watt bulb.
   4) Light from each of the 4 sources carries the same amount of information.
   5) Both 2 and 3 are equally valid as the best answer.

5. Suppose the student in Question 4 next illuminated a coleoptile from exactly opposite sides with a 100 watt bulb on each side. He observed that the coleoptile grew straight and did not bend toward either light. Using the data in Questions 4 and 5 which of the following would you most likely conclude?

   1) The coleoptile cannot perceive any information in light from a 100 watt bulb.
   2) Light from a 100 watt bulb contains no information.
   3) Light from each of the two 100 watt bulbs carries the same amount of information.
   4) Light from two 100 watt bulbs contains information equivalent to light from one 200 watt bulb.

5) Both 1 and 2 are equally likely as the best conclusion.

6. A biologist was watching an aquarium and observed a fish position itself so that another fish was directly facing the first fish. The first fish fully raised all its fins and vibrated its tail and fins, staying in front of the second fish. The biologist said, "The first fish is doing a courtship display to attract the second fish sexually." Which of the following would you say about the biologist and this statement?

1) He was accurately reporting his observation.
2) He was stating a testable hypothesis.
3) He was stating a conclusion which does not need to be tested.
4) He was making assumptions about the second fish.
5) He was wrong. The first fish was behaving aggressively to drive the second fish away from its territory.

7. A shark in a large tank is presented two wooden targets simultaneously 100 times every day. The targets are similar in every way except that one is blue and the other is red. The targets are switched frequently so that neither color is always to the shark's right or left. Each time the shark touches the red target he is given a piece of fish but he is given nothing for touching the blue target. The shark must choose one or the other target every time they are presented.

The data for the first week are:

Day	# Times Red Chosen	# Times Blue Chosen
1	48	52
2	50	50
3	59	41
4	60	40
5	72	28
6	90	10
7	98	2

Which of the following statements would be least acceptable?

1) Sharks prefer the color red over the color blue.
2) The shark could tell the difference between the two targets based on their color.
3) After a period of time we were able to "communicate" with the shark.
4) The shark "learned" that "red target" in some way meant fish.
5) Both 3 and 4.

8. In the mammalian kidney a rise in blood pressure is known to increase the total amount of urine excreted. A drop in blood pressure decreases the total amount of urine produced. A student proposes that this automatically regulates the animal's blood pressure. Which of the following is the student assuming?

1) A decrease in blood volume (total volume in the circulatory system) results in a decrease in blood pressure.
2) An increase in blood volume results in a decrease in blood pressure.
3) An increase in urine production results in a decrease in blood volume.
4) Both 1 and 3
5) Both 2 and 3

9. A student observed that when he held his breath, his heart rate would increase. He then stated that this was an example of a feedback mechanism. Which of the following was he assuming?

1) There are sensors which can detect how fast his heart is beating.

2) There are sensors which can detect the carbon dioxide concentration in the blood.
3) There are sensors which can detect the blood pressure in his arteries.
4) Both 1 and 2
5) All the above

10. Mechanisms exist in the body which maintain blood pressure at a relatively constant level. When blood pressure decreases, pressure receptors in the arteries cause nerve impulses to be sent to the muscle cells in the walls of the arteries. When pressure increases, the nerve impulses cease. Which of the following is most likely occurring?

1) Nerve impulses cause muscle cells to contract; when no impulses are received the muscle cells relax.
2) When pressure increases, the muscle cells are stimulated to contract thereby completing the negative feedback loop.
3) Nerve impulses can stimulate muscle cells to either contract or relax depending on the information concerning pressure.
4) When pressure decreases, the muscle cells receive the information to relax.
5) Both 2 and 4 are most likely occurring.

11. When blood flow only through the arteries to the kidney decreases, the systemic blood pressure in humans will increase. Decreased flow to the kidneys has no appreciable effect on blood flow to other parts of the body until after the blood pressure increases. A physiologist hypothesized that this observation indicates the presence of a control mechanism to insure that there is an adequate blood flow to the kidney. Which of the following is the physiologist most likely assuming?

1) The kidney can sense a decrease in blood flow.
2) Information on blood flow is transmitted from the kidney to the brain.
3) Information on blood flow can be transmitted from the brain to the kidney.
4) The kidney can secrete a hormone which increases blood pressure.
5) Choices 2 and 4 are equally valid as the most likely assumption.

12. From previous experiments a respiratory physiologist knew that a part of the brain (breathing center) can stimulate muscles which cause the chest to expand and air to be drawn into the lungs. He also knew that stretch receptors in the lungs could sense when the lungs were fully expanded and would then send nervous impulses to the breathing center. Which of the following must be occurring in this breathing control mechanism?

1) The stretch receptor must also send nervous impulses to the breathing center when the lungs are fully deflated.
2) The stretch receptors must also send nervous impulses to the muscles controlling lung expansion.
3) Stretch receptors must be present in both lungs.
4) The nervous impulses from the stretch receptor cause the breathing center to stop stimulating the muscles.
5) Nervous impulses from the breathing center must be sent to the stretch receptors.

13. Suppose the physiologist in Question 12 prevented the stretch receptors from sending nerve impulses to the breathing center. Which of the following would most likely occur?

1) If the lungs were already expanded, they would deflate.
2) The lung would expand and remain expanded.
3) The normal cycle of inhaling and exhaling would continue as usual.
4) The lungs would deflate and remain deflated.
5) Any air in the lungs would be forcibly exhaled.

14. When a whale comes to the surface of the water it exhales a stream of gas with great force through the blow hole in the top of its head. A scientist proposed that the whale's brain only sends impulses to the muscles which cause exhalation when the front part of the whale's head emerges from the water. Which of the following assumptions was the scientist most likely making?

1) The whales can perceive light underwater so that they know their distance from the surface.
2) When exposed to air, receptors in the skin of the whale's head send impulses to the brain.
3) Receptors in the whale's flipper can perceive a decrease in pressure as the whale nears the water's surface.
4) Whales inhale through their mouths and must exhale through their blowholes.
5) When exposed to air, the whale's muscles contract automatically.

15. A student placed a clamp on the carotid artery of a dog. The result was an increase in blood pressure when measured in the femoral artery. He proposed that information was transmitted from the carotid artery to the brain, then to the heart and other arteries, leading to the increase in blood pressure. He proposed that information could be transmitted from the brain to the heart and arteries by nerves or by a hormone. Which of the following experiments would distinguish between these two possibilities, assuming only one takes place?

1) Remove blood from the dog, transfuse it to a second dog with its carotid not clamped, and monitor blood pressure.
2) Clamp the second carotid artery and re-measure blood pressure.
3) Clamp the femoral artery and measure blood pressure in the carotid artery.
4) Inject a drug known to increase blood pressure, and determine if the same increase in pressure occurs as when the carotid is clamped.
5) Apply direct electrical stimulation to the dog's heart and monitor any change in blood pressure in the femoral artery.

16. Gastrin is a hormone which causes the stomach to release gastric juice. Cholecystokinin is a hormone which causes the gall bladder to release bile. Assume artery S supplies blood to the stomach and artery G supplies blood to the gall bladder. Which of the following is the best prediction and explanation?

1) If the arteries supplying the stomach and gall bladder are surgically switched so that artery G supplies the stomach and artery S supplies the gall bladder, then the stomach will release bile and the gall bladder will release gastric juice.
2) If the arteries supplying the stomach and gall bladder are switched the gall bladder will no longer release any bile because the artery going to it will be carrying the wrong hormone.
3) Switching arteries would have no effect because hormones are not transmitted by the circulatory system.
4) If the arteries supplying the stomach and gall bladder are switched the gall bladder will still release bile because it will receive cholecystokinin through artery S.
5) 2 and 3 are equally good predictions.

17. Blood from the small intestine and pancreas is transported to the liver by a blood vessel called the hepatic portal vein. The following data were obtained on blood glucose concentrations in normal and diabetic individuals.

	Entering liver via hepatic portal vein	Leaving liver
Normal	120	90
Diabetic	120	120

Which of the following statements is contradicted by the data?

1) In normal individuals the liver stores glucose as glycogen.
2) The diabetic absorbs less glucose from the small intestine than a normal individual.
3) In normal individuals increased glucose in the diet results in increased production of insulin.
4) Injection of insulin into the circulatory system of the diabetic results in a lower glucose concentration in blood leaving the liver than in blood entering.
5) Insulin is a hormone produced by the pancreas.

18.   Questions 18, 19 and 20 form a series. Each includes the information from the previous members of the series. In the green anole lizard, reproductive behavior of males is maintained by hormone T, secreted by the testes. If the testes are removed, reproductive behavior ceases. If hormone T is then injected, reproductive behavior resumes. A biologist made the hypothesis that hormone T acted on the brain at site A. He removed the testes from some lizards and inserted a small pellet of hormone T into site A in the brains of these animals. Reproductive behavior resumed. He concluded that his hypothesis was supported. Which of the following assumptions was he most likely making?

1) Hormone T would also act on the female brain.
2) Hormone T is converted to hormone E in the brain.
3) Hormone T is carried in the blood to some other site of action.
4) Hormone T is not carried in the blood away from the area where the pellet was placed.
5) Hormone T does not penetrate into the cells of the brain.

19.   Another biologist made the hypothesis that hormone T does not act directly on the brain, but that hormone E does. He showed that pellets of hormone E in the brain would also promote resumption of male reproductive behavior after the testes were removed. Which of the following observations would best support his hypothesis?

1) There is an enzyme in the brain that converts hormone T to hormone E.
2) There is an enzyme in the brain that converts hormone E to hormone T.
3) When radioactive hormone T is injected, radioactivity is found in brain cells.
4) Electrical stimulation of the brain results in reproductive behavior.
5) Electrical stimulation of the brain does not result in reproductive behavior.

20.   A third biologist was studying the effects of T on behavior of lizards without testes. When he injected drug D and hormone T simultaneously no reproductive behavior occurred. Which of the following experiments must he do to reach the conclusion that drug D acts by blocking the conversion of T to E?

1) Inject hormone T and E simultaneously.
2) Inject hormone E and drug D simultaneously.
3) Inject drug D into an animal with testes undisturbed.
4) Inject all three substances simultaneously.
5) Inject drug D alone.

21.   The tip was removed from a coleoptile. A substance was collected which diffused from coleoptile tips into gelatin blocks. This substance (auxin) was applied uniformly to the cut surface of the coleoptile where the tip had been removed. If this coleoptile is now illuminated from one side, which of the following would you most likely predict?

1) The coleoptile would bend toward the light source.
2) Growth in length would occur but not bending toward a light source.
3) The coleoptile would bend away from a light source.
4) There would be no growth in length and no bending.
5) From the data in this problem no decision can be made about which of the above choices is best.

22. An investigator exposed growing shoots to varying intensities of light. After exposure to the light the shoots were kept in the dark for 30-40 minutes. Then the coleoptile tips were cut off and kept for 45 minutes on agar. The agar was cut into blocks and each block was placed at the side of a cut shoot. The degree of curvature produced in these cut shoots was then measured. The results of three experiments are as follows:

Light	Degree of Curvature		
Intensity	Expt. I	Expt. II	Expt. III
0	11.5	11	8
1,000	6	5	4.5
10,000	Not done	Not done	11
100,000	15.5	16	10
1,000,000	None done	None done	14

Which of the following statements is most likely the hypothesis being tested by this experiment?

1) Light information induces production of a growth regulating substance in coleoptile tips.
2) Quantitative light information is chemically transmitted to the cut shoot.
3) The transmission of light information to the cut shoot is dependent on the time the shoots are kept in the dark and the time the cut coleoptiles are in contact with the agar.
4) Chemical information in the agar blocks results in curvature away from the block.
5) If something can go wrong with an experiment it will go wrong (Murphy's Law).

23. Several students observed the bending of coleoptiles toward a light source and proposed the following hypotheses regarding the hormone produced by the coleoptile tip.

A. Light can destroy the hormone.
B. The hormone migrates away from the light.
C. Light inhibits synthesis of the hormone in the tip.

If any one of these three hypotheses is valid, which of the following assumptions is necessary to explain the bending response?

1) When the coleoptile tip is removed, growth and bending are stopped.
2) No bending is observed when the coleoptile is kept in the dark.
3) Light can destroy the hormone.
4) The greater the hormone concentration, the greater the rate of cell growth.
5) Both 1 and 2

24. Suppose hypothesis B in the previous question is valid. Which of the following would you predict if the coleoptile is lighted equally from all sides?

1) It would grow straight.
2) It would bend in one direction, toward one of the lights.
3) No growth would be observed.
4) The coleoptile would grow outwards, i.e., it would get "fatter".
5) Either 1 or 3.

25. Suppose light stimulates the production of the hormone by the coleoptile tip. Which of the following would you predict?

1) It would bend away from light.
2) No bending would be observed when lighted from one side.
3) If the tip is removed, bending would still occur.
4) When the tip is covered with a black cap, bending will still occur.
5) It would bend toward light.

26. A nerve impulse occurs and travels along a nerve cell because of temporary changes in permeability to Na^+ and K^+ at the nerve cell membrane. A small amount of K^+ diffuses out of the cell and a small amount of Na^+ into the cell. Which of the following would prevent a nerve impulse?

1) Increase external Na^+.
2) Increase internal Na^+.
3) Increase external K^+.
4) Both 1 and 3
5) Both 2 and 3

27. A student proposed a hypothesis about nerve impulse transmission. In her hypothesis the information in an electrical signal is transmitted between nerve A and nerve B by a chemical A. This chemical transmitter is stored in vesicles only at the end of nerve A, and is released when the electrical signal reaches the end of nerve A. Chemical A then diffuses across the synapse gap and generates a new impulse in nerve B. Which of the following observations is *not* consistent with her model?

1) Addition of chemical A to the synapse results in the stimulation of nerve B.
2) Addition of a drug to the synapse inhibits the release of chemical A such that an impulse is not generated in nerve B.
3) Addition of chemical A to the synapse can lower the threshold of nerve B.
4) Lowering the temperature of the environment of nerves A and B increases the time required to transmit information from nerve A to B.
5) An electrical impulse can be transmitted from nerve B to A.

28. The transmitter substance released by the first nerve cell at a synapse is destroyed after it combines with receptors on the membrane of the second nerve cell. Which of the following would most likely occur if the transmitter substance were NOT destroyed?

1) A nerve impulse would not be transmitted from the first nerve cell to the second nerve cell across the synapse.
2) Few, if any, nerve impulses would be measured at the first nerve cell.
3) The second nerve cell would continue to generate nerve impulses in the absence of impulses from the first nerve.
4) No more transmitter substances would be produced by the first nerve cell.
5) Both 2 and 3.

29. Which of the following would prevent the transmission of a nerve impulse at a synapse?

1) Inhibition of breakdown of transmitter substance after combination with receptor sites on the second nerve cell.
2) Inhibition of transmitter release by the first nerve cell.
3) Narrowing the gap between the two nerve cells at the synapse.
4) Inhibition of the diffusion of transmitter substance between the two nerve cells.
5) Both 2 and 4.

30. Consider a chemical, A, which is known to exist in the vesicles in the synapses at the cat brain. If A is applied directly to brain cells, nerve impulses are the result. However, if A is injected into the cat's circulatory system, it has no effect on the brain. Which of the following is the least acceptable explanation of the observations?

1) There are enzymes in the blood which destroy A.
2) The blood vessels in the brain are specially constructed to prevent the passage of chemicals out of the blood onto the brain cells.
3) When an injection of A is mixed into the blood, it becomes too dilute to be effective.

4) A is an inhibitor of nerve impulses.
5) Both 1 and 3 are unacceptable explanations.

31. Cigarette smoking increases the amount of nicotine in a person's body. Nicotine can mimic the action of the chemical found in some synaptic vesicles. From this information, which of the following is the best prediction of what happens in a person who smokes 20 cigarettes/day compared to a non-smoker? Assume other activities are equal.

1) The non-smoker probably has more total nerve cells firing.
2) The smoker is a less nervous person.
3) The smoker probably has more total nerve cells firing.
4) The non-smoker is deficient in nicotine and thus has slower reflexes.
5) There is no difference between the smoker and the non-smoker in the number of nerve cells firing.

The following information should be used in answering the next two questions. In tropical Central and South America there are two closely related species of butterflies that look almost exactly alike. They can only be distinguished by looking at the underside of the front wing of the male, near where it joins the body. There the males of one species have four red dots, which are missing in the other species. The females look exactly alike and lack dots. During courtship the males of both species hover in front of the females and do a looping, aerial dance. If the female who is thus courted is ready for mating, she lands on a bit of vegetation and folds her wings. He lands beside her and tilts his folded wings away, showing their undersides. She either stays there (whereupon copulation occurs) or she leaves, signaling rejection.

32. A scientist hypothesized that the four red dots on the males of the one species were used by the females in choosing mates. In testing his hypothesis, he reasoned that the butterflies would have to be able to see colors (have color vision) if they were to choose on the basis of the red dots. He therefore put electrodes on the optic nerves of the butterflies and recorded the nerves' responses to flashes of colored light. His results are given below in the form of oscilloscope tracings. In each case a different nerve cell was being recorded, but the sequence of colored light flashes in front of the butterfly's eye was the same: first blue, then green, then yellow, then orange, then red.

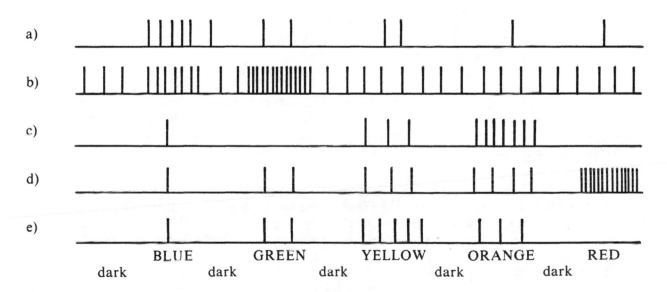

Which of the nerve cells coming from the eye would a female butterfly be most likely to "pay attention to" when choosing a mate?

1) cell (a)
2) cell (b)
3) cell (c)
4) cell (d)
5) cell (e)

33. If the scientist's hypothesis is correct, how would the males choose among the females?

    1) They would court only females whose red dots matched their own.
    2) The males with red dots would choose females lacking them, and males without red dots would choose females with them.
    3) The males would not be able to choose at all, but would court females of both species.
    4) The males would sit around on vegetation waiting to check out females for red dots.
    5) The males would seek out females of other species as mates.

34. A physiologist was investigating the function of a pressure sensor in the foot of a tree climbing organism. The pressure sensor is a microscopic structure in the skin. Nerve cells transmit information from the sensor to other parts of the body. While making electrical measurements from these nerves, the physiologist observed that no action potentials occurred until a pressure of 5 lbs/sq. in. was applied. As the pressure was slowly increased up to 15 lbs./sq. in. the frequency of the action potentials did not change. Which of the following would be the best conclusion?

    1) The organism cannot detect pressure differences between 3 and 7 lbs/sq. in.
    2) The organism cannot detect pressure differences between 8 and 20 lbs/sq. in.
    3) The organism cannot detect pressure differences between 7 and 12 lbs/sq. in.
    4) The organism can detect differences between 6 and 13 lbs/sq. in.
    5) Both 1 and 2 are equally likely as the best conclusion.

35. The physiologist in the previous question quickly decreased the pressure applied to the sensor from 15 lbs/sq. in. to 10 lbs/sq. in. while observing action potentials in the nerve fibers. He found that immediately after the rapid pressure change the frequency of the action potentials increased for 1.5 seconds and then decreased to the level previously observed. Which of the following could you most likely conclude from the data and observations in Questions 34 and 35?

    1) The frequency of individual electrical changes indicates the amount of pressure applied.
    2) The organism can detect rapid pressure changes between 13 and 8 lbs/sq. in.
    3) The organism can detect rapid pressure changes between 4 and 0 lbs/sq. in.
    4) Rapid pressure changes cannot be detected by this organism.
    5) Both 2 and 3 are equally valid as the most likely conclusion.

36. A chemical substance transmits information from nerves to muscles in structures very much like synapses. An enzyme present in the gaps breaks down the substance and allows the muscles to relax. Certain poisons cause paralysis by inhibiting this enzyme. Since the chemical transmitter is not destroyed, the muscle contracts but cannot relax. How would you treat a patient poisoned in this way?

    1) Forcefully move his arms and legs to stretch the muscles.
    2) Feed him by mouth a dose of the enzyme that breaks down the transmitter.
    3) Inject a drug that stimulates transmission of nerve impulses.
    4) Inject a drug that blocks some of the receptors on the muscle cell which detect the transmitter substance.
    5) Remove the contents from his stomach.

37. A zoologist was observing the response of a single muscle in the leg of an organism when two nerves, A and B, were stimulated at distant parts of the body. He made the following observations:

A. When nerve A alone is stimulated the muscle contracts.
B. When nerve B alone is stimulated the muscle does not contract.
C. When nerves A and B are stimulated simultaneously, the muscle does not contract.

Which of the following is the best interpretation of these observations?

1) Nerves A and B lead directly to the muscle via separate pathways.
2) Nerve B transmits no information in this system.
3) Nerves A and B synapse on one nerve leading to the muscle.
4) Nerve A synapses on a nerve leading to the muscle; nerve B leads directly to the muscle.
5) Nerve B synapses on a nerve leading to the muscle, nerve A leads directly to the muscle.

38. The variation of contraction force with the strength of electrical stimulation in a single muscle fiber and in a whole muscle are very different. The following assumptions might be made regarding these data:

A. Different muscle fibers have different thresholds.
B. Different muscle fibers exhibit different forces of contraction when stimulated.
C. The strength of an electrical stimulation decreases as it spreads through a whole muscle.

Which of the above assumptions must necessarily be made to remove any conflict between the data on single muscle fibers and whole muscle?

1) A and B
2) A, B, and C
3) B or C
4) B and C
5) A or C

39. A physiologist electrically stimulated the nerves leading from stretch receptors in the muscle. She maintained this stimulation over a long period of time. Which of the following would most likely occur?

1) The muscle would show sustained contraction.
2) The stretch receptors would first contract and then relax.
3) Nerves leading to the muscle would not transmit electrical impulses.
4) Both 1 and 2.
5) All the above.

40. In one frog a spindle fiber (stretch receptor) of a leg muscle was stretched, resulting in a contraction of the entire muscle. However, the stretch receptor was prevented from returning to its "prestretch" position when the muscle contracted. Which of the following best explains the comparison of this muscle to a normal muscle in which the receptor shortens when the muscle contracts?

1) The contraction of the muscle with the continually stretched receptor would be less because the stretch receptor would become exhausted.
2) The contraction of the muscle with the continually stretched receptor would be less because muscle contraction is linearly related to the amount of relaxation of the stretch receptor.
3) The contraction of both muscles would be the same because the position of the stretch receptor doesn't matter once it initiates an impulse.
4) The contraction of the muscle with the continually stretched receptor would be greater because the receptor would continue to initiate impulses which would result in repeated contraction of the muscle fibers.

5) The contraction of the muscle with the continually stretched receptor would be greater because for every action of muscle fiber there is an opposite reaction of another fiber.

41. A physiologist was investigating the function of a light receptor in an organism. Nerve cells lead from this receptor to other parts of the body. The physiologist made electrical measurements in those nerves while gradually increasing the light intensity on the receptors. The results are shown in the graph below.

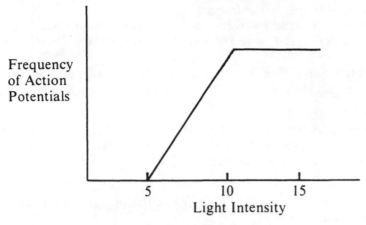

Which of the following would be the best conclusion?

1) The organism cannot detect the difference between 3 and 7 light intensity units.
2) The organism can detect the difference between 6 and 9 light intensity units.
3) The organism cannot detect the difference between 11 and 18 light intensity units.
4) The organism can detect the difference between 9 and 12 light intensity units.
5) Both 2 and 4 are valid conclusions.

42. The physiologist then studied a small muscle which received impulses carried by the nerve in Question 41. He made the following observations:

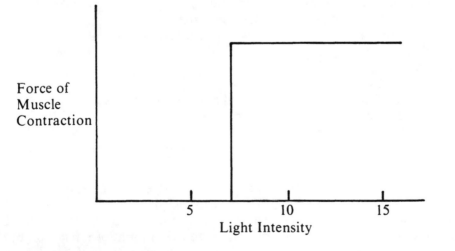

Which of the following is most likely true of this muscle?

1) All the fibers in this muscle have the same threshold.
2) The force of contraction will be greater at light intensity 20.
3) All the fibers in this muscle receive the same stimulation from the nerve.
4) Several nerves conduct impulses to this muscle.
5) Both 1 and 3 are equally likely.

43. The muscle in Question 42 is attached to an organ of locomotion.. The animal only moves about when this muscle contracts. An animal behavioralist, who knew nothing about the data presented in Questions 41 and 42 (You do know about these data) studied the responses of this animal to variations in light intensity over the range 0-15 intensity units. Which of the following would the behavioralist most likely conclude?

1) The organism cannot detect the difference between 3 and 7 light intensity units.
2) The organism can detect the difference between 6 and 9 light intensity units.
3) The organism cannot detect the difference between 11 and 18 light intensity units.
4) The organism can detect the difference between 9 and 12 light intensity units.
5) Both 2 and 4 are valid conclusions.

44. The behavioralist in Question 43 concluded that there must be nerves carrying information from the light receptor to the locomotion muscle in this organism. He made the hypothesis that other nerves arrive at this muscle, carrying information from other receptors. Which of the following observations would support this hypothesis?

1) The animal stops moving when the light intensity drops suddenly from 8 to 0.
2) The animal continues to move when a bell is rung while the light intensity is 11.
3) The animal remains still when the light intensity is 4 and ammonia gas is released into the environment.
4) The animal continues to move when the light intensity drops suddenly from 8 to 0 while a bell is rung.
5) Flickering lights also cause the animal to move.

45. The Mantua is an imaginary aquatic animal. The males swim about but the females remain attached to rocks in small caves, thus the male must receive some signal from the female in order to locate her for mating. A student proposed two hypotheses:

1) The female emits a chemical signal which the male detects.
2) The female emits sounds which the male detects.

To test these hypotheses, he placed a male Mantua before two caves. In one cave a female Mantua was present. In the other cave several female Mantuas had been living but were removed just before the male was tested. The male entered the cave where the female was present and the student concluded that the signal for the female was sound. What was he most likely assuming?

1) The female Mantua does not emit any chemicals.
2) The male Mantua cannot perceive any chemicals emitted by the female.
3) Chemicals emitted by the female Mantua persist in the water at least as long as the duration of the experiment.
4) The female Mantua perceives some signal from the male.
5) He made no assumptions in order to reach his conclusion.

46.   The oviducts of the adult female ringdove will increase 500% in size when she is placed in a cage with a normal male. The increase also occurs if the two are placed in adjacent rooms with a window between them. It does not occur if the male is castrated. Which of the following adequately explains this?

   1)   The female is translating olfactory (smell) inputs into a hormonal response.
   2)   The female is translating visual inputs into a hormonal response.
   3)   The castrated male does not appear the same as a normal male.
   4)   Both 2 and 3 would explain the situation.
   5)   None of the above offer a good explanation.

PHYSIOLOGICAL ECOLOGY

The organisms in an ecosystem can be visualized as occupying the different trophic levels of producers (usually green plants), herbivores, carnivores, and decomposers.

An ecosystem is characterized by nutrient cycling within the system and by energy flow through the system. Energy usually enters the system as radiant energy and is captured by photosynthesis. Not all radiant energy is captured and some is lost by respiration. The net productivity of the producers is thus defined as photosynthesis minus respiration. The energy in a trophic level at any time is stored in the biomass of the organisms.

In the process of succession, an ecosystem may change over time due to the interactions of organisms with each other and with the physical environment. Patterns of succession may be understood by examining these interactions in detail.

Interactions within and between populations can be revealed by changes in population size with time. Survivorship curves can be interpreted in terms of interaction factors such as degree of parental care and predation and individual factors such as life span, reproduction rates and physiological responses to environmental variables.

EXAMPLE 1.

An investigator placed 5 different kinds of organisms in a sealed chamber and established all conditions (such as light, temperature, water, etc.) necessary to maintain them for an extended period of time. One of the organisms was known to be a plant in which the starch had been previously labeled with radioactive carbon and one was known to be a carnivore. A student hypothesized that ultimately significant radioactive carbon would be found in all the organisms. What was the student most likely assuming?

    1)   The other three organisms were carnivores.
    2)   At least one of the other organisms was an herbivore.
    3)   The plant with radioactive carbon would use some of its starch in respiration.
    4)   At least two of the other organisms were green plants.
    5)   The carnivore respired more rapidly than the plant.

ANALYSIS:

It is necessary first to interpret the phrase "significant radioactive carbon." One reasonable interpretation is carbon incorporated into the biomass of the organism as opposed to CO_2 which had simply diffused into the body fluids of the organism.

There are two pathways by which C^* can go from starch in a plant into the biomass of another organism. First, the plant can metabolize some starch, forming C^*O_2, which is then taken up by another plant and incorporated into starch by photosynthesis. Second, the plant may be eaten by an herbivore, resulting in incorporation of C^* into herbivore biomass. From here it may be released as C^*O_2 and find its way into other plants, or it may pass directly into a carnivore that eats the herbivore.

In this problem, the hypothesis is that C^* moves from the plant into at least one carnivore. The only way for this to occur is by the plant being eaten by an herbivore which is eaten in turn by the carnivore. Thus he is assuming that there is at least one herbivore in the system (Choice 2). This argument eliminates Choice 1. Choice 5 is irrelevant because respiration is not involved in the transfer mechanism. Choices 3 and 4 are possible but not necessary.

If you consider C^*O_2 in the body fluids to be "significant radioactive carbon" and if you state this assumption, then you would most likely also assume that the plant uses some of the starch in respiration and releases C^*O_2 which enters other organisms by diffusion. This is the most direct pathway, although all the other assumptions are also possible.

EXAMPLE 2.

When a certain organism is grown in the laboratory, its population increases for a certain time, then becomes steady. A scientist made the hypothesis that limited availability of food caused the limited population growth. Which of the following observations would provide the strongest support for this hypothesis?

    1)   When population growth ceases, there is still food available to the organisms.
    2)   When less food is made available to the organisms at the beginning of the experiment, lower steady populations are achieved.
    3)   A lower steady population of the original organism is reached when a similar organism is also present in the same culture vessel.
    4)   The same steady population is achieved when the original study is repeated.
    5)   A higher steady population is achieved when chemical X is sprayed into the culture vessel daily.

ANALYSIS:

There are at least two ways to think of this situation. The simplest is to assume that the observations of steady population size were made for a short time and that by "limited availability of food" the scientist means that the culture has exhausted its food supply. In this case the birth and death rates are both equal to zero temporarily, but at some time death from starvation will begin and the population size will drop to nothing. A more complex interpretation is to assume that food is still available to the organisms, but in a limited amount in each time period. In this case the food supply would be just adequate to allow survival of a certain number of individuals and to support reproduction at a rate equal to the death rate.

1.   If we observed food available to the organisms during the period of steady population size, we could not make the first interpretation. Those assumptions include the idea that food supply is exhausted. This observation would not support that hypothesis.

This observation would be consistent with the second interpretation of the hypothesis because that interpretation does involve continued food availability. However, this is not support of the hypothesis. It does not eliminate any other possibilities such as accumulation of toxic waste.

2.   This observation would provide some support for the hypothesis, especially for the first interpretation of it, because it provides a direct link between food supply and population size, as the hypothesis proposes. Note, however, that this is not conclusive evidence. For example, it is also consistent with the hypothesis that population size is limited by accumulation of toxic waste products since these products are formed from food and would be proportional to food supply.

3.   This observation adds another variable to the system and makes it even more difficult to reach a conclusion. If the two organisms compete for food, their food could be an even more stringent limit on population size. On the other hand, the presence of other organisms could also increase the rate of toxic waste formation or reduce population size by some other interaction.

4.   This observation would have no effect on the hypothesis. It would increase our confidence that the observations were valid and not some fluke event, but all of the possible interpretations of the observations are still possible.

5.   As with 3 above this observation involves addition of another variable to the system which further confounds analysis. This observation is consistent with the scientist's hypothesis (Chemical X increases the efficiency with which the organisms utilize their food - it's a sort of dietary supplement) or with others (Chemical X neutralizes toxic waste products).

1. Algae were grown in two aquariums - aquarium A without herbivores and aquarium B with herbivores. The dry weights of living algae and herbivores were determined at the start of the experiment (day 1) and again five days later (day 6); weights of dead algae and herbivores were ignored. The results of this experiment are given in the following table:

Day	Aquarium A Algae (g)	Aquarium B Algae (g)	Aquarium B Herbivore (g)
1	0.5	0.5	0.1
6	0.7	0.6	0.2

Which of the following is the best interpretation?

1) The algae produced the same amount of biomass in aquarium A and B.
2) The algae produced more biomass in aquarium B, but most of it was eaten by the herbivores.
3) The algae produced more biomass in aquarium A, but most of the algae died.
4) The data cannot be correct as presented.
5) The data may be correct as presented, but they do not show in which aquarium the algae produced the most biomass.

2. Two biology students attempted to set up a stable freshwater "mini-ecosystem" in an aquarium filled with unfiltered pond water. They added green plants, herbivores, small carnivores and large carnivores. After several weeks, they found no herbivores. Which of the following is the best explanation of their observation?

1) There were too many herbivores in the aquarium.
2) There were too many small carnivores in the aquarium.
3) There were too many large carnivores in the aquarium.
4) Both 1 and 3.
5) Both 2 and 3.

3. A biologist placed 20 insects, 10 spiders, and 50 green plants into a closed chamber. He predicted that after 2 months the numbers of the enclosed living organisms would not change significantly. Which of the following did the biologist NOT assume when making his prediction?

1) The spiders are carnivorous.
2) The plants contain enough energy to sustain the insects and spiders.
3) The insects and/or the spiders are herbivorous.
4) Reproduction of organisms will equal the loss of organisms by death and predation.
5) The environmental conditions inside the chamber are adequate to sustain life in all the organisms.

4. A student constructed an ecosystem by planting an aquatic plant *(Elodea)* in an aquarium. Subsequently he added a few snails and some fish. He then sealed the aquarium with a glass plate and placed it next to a window in his room. After several months he observed that the living organisms in the aquarium were surviving and apparently doing well. Which of the following statements about the aquarium is least likely?

1) The plants are the producers in this system.
2) No energy gets into the aquarium from the outside.
3) Carbon dioxide is produced in the aquarium and utilized by the *Elodea*.
4) Energy is transferred from one organism to the other.
5) Herbivores utilize both the *Elodea* and the oxygen produced by *Elodea*.

5. A student observed over a period of several days that the level of oxygen in a sealed aquarium

was twice as great during the day as at night. It was known that the aquarium contained only algae. He then made the statement that algae used 1/2 as much oxygen in respiration during the day as at night. This student knew that the algae were photosynthesizing during the day. Which of the following would be most justified regarding his statement:

1) The data support his statement.
2) His statement can be criticized because he has neglected the oxygen.
3) The data contradict his statement.
4) His statement can be criticized because he does not distinguish between aerobic and anaerobic glucose breakdown.
5) The data neither support nor contradict his statement.

6. Carbon dioxide containing radioactive carbon was bubbled into an illuminated aquarium containing unfiltered pond water. In what order will radioactivity appear in the various organisms in the aquarium?

1) Carnivores - herbivores - producers
2) Herbivores - carnivores - producers
3) Producers - herbivores - carnivores
4) Producers - carnivores - herbivores
5) Herbivores - producers - carnivores

7. An aquarium was established with plants and bacteria. The aquarium was sealed to prevent the gain or loss of matter, but energy (light and heat) could enter or leave. After an initial period of growth the biomass of plants and bacteria stayed nearly constant, even though the plants continuously lost old leaves and grew new ones. A very small amount of radioactive carbon (^{14}C) was added to the aquarium as $^{14}CO_2$. A student predicted that equal proportions of ^{14}C eventually would be found in live plants and live bacteria. Which of the following must the student assume to make her prediction?

1) An insignificant amount of radioactive carbon (^{14}C) changes to non-radioactive carbon (^{12}C).
2) Bacteria take up $^{14}CO_2$ and incorporate it into their biomass.
3) None of the ^{14}C will become incorporated into dead plants and dead bacteria.
4) ^{14}C and ^{12}C are not distinguished by the organisms, and the proportion of ^{12}C is the same in live plants and live bacteria.
5) The student must assume all of the above to make her prediction.

8. Chipmunk A was fed some glucose labeled with radioactive carbon while chipmunk B was fed only unlabeled glucose. Then both chipmunks were placed in a sealed, lighted growth chamber containing several green plants. In which of the following cases will radioactive carbon most likely appear in molecules synthesized by chipmunk B?

1) Chipmunk B eats some of the plants at a later time.
2) Both chipmunks exercise vigorously for an extended period.
3) Chipmunk A eats some of the plants at a later time.
4) The oxygen concentration in the chamber is increased to twice the normal level.
5) Both 1 and 3 are equally valid as the best answer.

9. Which of the following would INCREASE the time required for radioactive carbon to appear in chipmunk B in Question 8?

1) Increase the size of the chamber.
2) Increase the light intensity.
3) Decrease the respiration rate of chipmunk A.
4) Decrease the size of the chamber.

5) Both 1 and 3 are equally valid as the best answer.

10. A biologist was studying the incorporation of nitrogen into the biomass of a plant. He wanted to test the hypothesis that nitrogen from the air entered the plant only through the leaves and became fixed into the plant's molecules. He exposed a plant, growing in a pot of soil and well lighted, to an atmosphere containing labeled nitrogen atoms for eight hours. At the end of this period he isolated from the leaves some molecules containing labeled nitrogen. He claimed that these observations supported his hypothesis. Which of the following is the best criticism of his claim?

1) The whole plant was exposed to labeled nitrogen, not just the leaves.
2) Eight hours is not long enough.
3) He assumed that the soil contained labeled nitrogen atoms.
4) He should have stripped all the leaves from the plant before the experiment.
5) He should have used only detached leaves in the experiment.

11. The same biologist then repeated the experiment using several specimens of the same kind of plant. He analyzed the leaves, the stems and the roots separately for labeled compounds. He found labeled compounds in the roots after one hour, in the roots and stems after four hours, and in all three parts after eight hours. Using all his observations, which of the following conclusions is most justified?

1) Nitrogen enters the leaves of the plant and is fixed in organic molecules.
2) Nitrogen enters the leaves and is distributed throughout the plant.
3) Nitrogen enters the roots of the plant and becomes fixed in organic molecules which are distributed throughout the plant.
4) Nitrogen enters the roots, is distributed to all parts of the plant and becomes fixed in organic molecules in all parts.
5) 3 and 4 are both justified by the evidence.

12. It is known that nitrogen moves from organism Z to organisms X and Y. These are the only losses of nitrogen from organism Z. The following data were collected:

NITROGEN CONTENT (mg)

ORGANISM	DAY 1	DAY 14
X	15	18
Y	5	7
Z	10	10

How can these observations best be explained?

1) The movement of nitrogen from Z to X is greater than from Z to Y.
2) The movement of nitrogen is greater from X to Y than from Z to Y.
3) Nitrogen is moving into and out of organism Z.
4) The movement of nitrogen is greater from Z to Y than from Z to X.
5) Both 1 and 3 are valid explanations.

13. Which of the following assumptions did you make in Question 12?

1) The loss of nitrogen from X is constant.
2) The loss of nitrogen from X and the loss of nitrogen from Y are equal.
3) More nitrogen is lost from X than is lost from Y.
4) The total loss of nitrogen from Z is not constant.
5) Both 1 and 4 were assumed.

14. An aquarium is filled with sterilized water. Into the aquarium are placed 2 fish and several

green plants. The aquarium is sealed and left for several months. One of the fish dies. What is the best prediction of what will then occur in the aquarium?

1) The elements in the dead fish will become available for the plants to utilize.
2) The organic molecules in the dead fish will be released and absorbed by the plants.
3) The dead fish will decompose rapidly.
4) The elements in the dead fish will not be recycled because the fish will not be decomposed.
5) The live fish and plants will cycle their elements more quickly.

15. Which of the following would contribute the most to an increase in the net productivity of a sealed aquarium?

1) Increase the rate of anaerobic respiration.
2) Increase the rate of CO_2 uptake by the plants.
3) Increase the duration of the daily dark period.
4) Increase the number of fish.
5) Increase all of the above.

16. A biologist was studying a small aquatic animal which fed only on one species of alga. During one three-month period he gathered the following data on weights (grams per cubic meter of water) of the animals and of the algae serving as a food source:

	Weight (g/m^3)		
	June	July	August
Animals	8	7	9
Algae	6	4	6

Which of the following would the biologist most likely infer from these data?

1) The animals must have some food source in addition to the algae.
2) The algae serve as a food source.
3) The algae reproduce and grow rapidly but are quickly consumed by the animals.
4) Measuring the weight of animals and algae in terms of grams per cubic meter is not a desirable method.
5) The animals reproduce and grow more rapidly than the algae.

17. In studying the total weights of organisms over a long period of time in an area, a scientist made the following observations of average values:

$$Carnivores \ 0.47 \ g/m^2$$
$$Herbivores \ 0.31 \ g/m^2$$
$$Plants \ 267 \ g/m^2$$

Which of the following is the best explanation of these observations?

1) Carnivores cannot weigh more than the herbivores over a long period of time, therefore carnivores will slowly die out until they are less than the herbivores.
2) Carnivores will eat all the herbivores, and the carnivores of the area will die out also.
3) The plants will begin producing more, so that the herbivores will become more numerous.
4) Herbivores reproduce much more quickly than carnivores and are replaced as rapidly as they are consumed.
5) This cannot be an actual situation, therefore, it can have only a theoretical solution.

18. A group of people decide that the agriculture and eating habits of the people in a certain area should be designed to support the maximum population on that area of land. Various vegetables, herbivores, and carnivores (such as wolves) will all thrive on this land. Assume the humans and herbivores are equally efficient at extracting energy from their food. Also assume protein and

vitamin requirements can be met by eating any of the above organisms, but each person requires a certain number of kilocalories per year. Which of the following policies should support the greatest density of humans?

1) Raise and eat only plants.
2) Eat only herbivores.
3) Raise and eat both plants and herbivores in approximately equal quantities.
4) Eat all of the organisms in the food chain.
5) Eat only the carnivores, but add large quantities of meat tenderizer to the wolves.

19. A student wanted to estimate the net productivity of the algae in a pond. He measured total photosynthesis in the pond as 11000 KCal and respiration as 7000 KCal during a 24 hour period. The net productivity would be:

1) 18,000 KCal
2) 4,000 KCal
3) 11,000/7,000 = 1.57 KCal
4) 77,000 KCal
5) Cannot be calculated from the above data.

20. A biologist observed that a certain insect eats nothing except the dung of herbivores. Which of the following would be the most justifiable interpretation?

1) As it passes through the digestive system of the herbivore, something having food value to insects is added to the food mass.
2) Herbivore dung has higher energy content than carnivore dung.
3) The digestive systems of herbivores do not extract all of the energy from the food eaten.
4) Plant eaters are more wasteful of food energy than meat eaters.
5) The digestive systems of herbivores do not extract any of the energy from the food eaten.

21. There was once a man who proposed to make money from a rat and cat farm. He would grow rats to feed the cats. He would sell cat furs for coats. He would feed cat carcasses to the rats. He figured he therefore needed only to buy for the rats a quantity of grain containing as much energy as the cat fur. Which of the following assumptions was he most likely making?

1) The KCal/gm in cat fur and grain are the same.
2) The energy content of rat feces is 0 KCal/gm.
3) All the energy absorbed by a cat from its food is converted to biomass.
4) All of the above.
5) Only 2 and 3 above.

22. An experimental fish was being pursued by a predator: its only means of escape was to flee. This fish was examined just prior to the incident mentioned above. The following observations were made:

Substance	Amount in Blood
Glucose	5
Oxygen	70
CO_2	43

It was also noted that various salts were present and the pH was 6.8. If you repeated these observations shortly after the "chase scene" began, which of the following would most likely be expected to decrease?

A. Glucose
B. Oxygen
C. CO_2
D. pH

1) A only.
2) A and B only.
3) A, B and C only.
4) A, B, and D only.
5) All of the above.

23. Sundew is a small green plant which has the ability to trap insects in a sticky substance secreted by its leaf cells. The insects are then digested by enzymes secreted by the plant. A biologist hypothesized that the sundew does not produce glucose by photosynthesis, but obtains all its energy from the molecules of trapped insects. Which of the following experiments would best test the biologist's hypothesis?

1) Grow the sundew in the dark with no insects available.
2) Grow the sundew in the light with insects available.
3) Grow the sundew in the dark with insects available.
4) Grow the sundew in the light with no insects.
5) All of the above together would be the best experiment.

24. A green pigment was extracted from a sundew plant. The absorption spectrum of the pigment was determined to be identical to that of chlorophyll found in other green plants. The biologist in Question 23 now altered his hypothesis. He now hypothesized that the sundew can and will produce glucose by photosynthesis but only when no insects from which to obtain glucose are available. He set up several sets of plants and grew them in the following conditions:

Set A: Atmosphere containing $^{14}CO_2$, light, insects available.
Set B: Atmosphere containing $^{14}CO_2$, light, no insects available.
Set C: Atmosphere containing $^{14}CO_2$, dark, insects available.
Set D: Atmosphere containing $^{14}CO_2$, dark, no insects available.

After 2 days tests were made to determine the amount of ^{14}C in leaves from plants in each set. If the biologist's new hypothesis is valid, which sets of leaves would most likely contain ^{14}C?

1) Set A only.
2) Set B only.
3) Sets C and D only.
4) Sets B and D only.
5) Sets A and B only.

25. For many years, it has been assumed that carnivorous plants used the nitrogen from the insects they "ate". Which of the following several tests would be best to determine the validity of this assumption?

1) A bomb calorimeter test might show whether the energy level per gram of the plants and the insects was similar; if so, the plants might have been using the insects.
2) By measuring the biomass of the plants in a particular area, and comparing it with the biomass of the animals of the area, it would be apparent whether the ratios were correct.
3) By "feeding" the plants insects containing radioactive nitrogen one could later measure whether the plants had radioactive nitrogen.
4) An insectivorous plant would get its nitrogen from the air, not from an insect, and these tests would not work.
5) A plant in light (as it would be in nature) would not respire, therefore none of the above would work.

26. Beach grass is commonly planted on coastal dunes to stabilize them and prevent sand movement. It has been found that the number of species and size of populations of burrowing arthropods was lower in dunes with beach grass than in dunes with other plants. One ecologist has

proposed that the dune grass roots form a dense mat which prevents movement by burrowing creatures. Which of the following observations would give the strongest support to this hypothesis?

1) Burrowing arthropods do not eat beach grass.
2) Burrowing arthropods die when exposed to sand in which beach grass was grown.
3) Burrowing arthropods are extremely rare in open sand with no plant growth at all.
4) Burrowing arthropod populations are very low in an area in which an unusually dense growth of a different plant occurs and the roots are thickly matted.
5) 1 and 3 are equally strong support.

27.   An ecology student proposed that plant succession could be totally explained by germination and growth rate. For example, crabgrass seed germinates and grows quickly and appears early in succession while acorns are slow to germinate and oak trees, which appear later, grow slowly. Which of the following is he most likely assuming?

1) Seed germination of plants in later stages would be more rapid in higher light intensities.
2) A greater proportion of crabgrass seed germinate than of acorns.
3) Larger plants in later stages require more water and nutrients.
4) Seeds from plants in all successional stages are present at the initial stage.
5) Both 2 and 3 are equally valid as the best choice.

28.   It has been observed that plants which grow in the late stages of succession in an abandoned field tend to have larger seeds than the plants which grow in the early stages. Which of the following statements is most pertinent to an explanation of this observation?

1) Large seeds are harder for seed-eating birds to crack.
2) Small seeds are harder for seed-eating birds to find.
3) Photosynthesis produces more water in the late stages of succession.
4) Photosynthesis proceeds at a slower rate in the shade.
5) Photosynthesis proceeds at a faster rate when the temperature is lower.

29.   It has been shown that seed size is correlated with germination and growth in varying light intensities. That is, small seeds germinate and grow better in high light intensity than in low, while large seeds are more successful in shade. It has also been observed that the succession of plants on abandoned fields is a trend from plants with small seeds to those with larger. For example, acorns are larger than pine nuts which are larger than huckleberry seeds. How can these observations aid in the explanation of succession?

1) Small seeds are generally more delicate and cannot withstand harsh physical conditions.
2) Large seeds are generally easier for squirrels to find and eat.
3) In succession the plants tend to be larger in the later stages, making surface light intensity lower and thereby favoring plants with larger seeds.
4) In succession, the later plants tend to produce more seeds, thereby compensating for their poorer growth.
5) In succession, the surface temperature decreases in later stages, therefore small seeds will be warmer in early stages and will grow faster.

In 1931 a biology student was studying a stable community of organisms living on a rock surface at the seashore. In November he scraped off all the organisms from one square yard of rock and counted the individuals. He observed this test square at intervals thereafter and counted but did not remove any organisms. His data, given below, will be used in the next three questions.

Organism	Nov.	Feb.	Mar.	Apr.	June	July	Oct.
Mussels	1612	0	0	55	51	7	53
Worms	327	0	0	0	0	0	0
Crabs	430	0	0	0	0	0	0
Isopods	926	0	0	0	0	0	0
Barnacles —							
Short species	950	0	518	2542	2345	2275	2534
Tall species	356	0	62	109	109	109	109
Limpets	329	62	239	322	364	342	1012
Purple snails	218	0	0	2	7	367	47

Mussels are clam-like creatures which permanently fasten themselves to the rock by means of tough fibers and which form dense masses covering the rock in areas where wave action is vigorous. Worms, crabs and isopods are small creatures which move about freely but cannot attach to or grip the rock surface. Barnacles live permanently attached to any convenient surface. Limpets and purple snails move about in search of food and can grip the surface firmly.

30.   Prior to this study biologists believed that succession does not occur on these rocks. This student concluded that his observations were evidence that succession does occur. Which of the following predictions would he most likely make?

1)   At some future time the original populations will be restored.
2)   The populations will be unchanged for the next ten years after his study.
3)   Random population changes will continue to occur.
4)   At some future time stable populations will be restored.
5)   Purple snails eat mussels.

31.   Assume that all of these organisms reproduce by releasing microscopic larvae into the water. These larvae float around in currents for 2-3 weeks, then settle on the surface of the rocks and develop adult forms. If the surface is suitable for their way of life they live; if not, they die. Larvae may be released in three seasons: Spring (March and April), Summer (June and July) and Fall (September). Which of the following organisms release larvae more than once a year?

1)   Mussels.
2)   Short barnacles.
3)   Tall barnacles.
4)   Purple snails.
5)   Both mussels and short barnacles.

32.   Which of the following hypotheses could best account for the drop in the mussel population in July?

1)   They were crowded out by barnacles.
2)   They were poisoned by a substance continuously released by limpets.
3)   They were eaten by baby purple snails.
4)   They starved because their only food source is worms.
5)   2 and 3 are equally good hypotheses.

33.   Sea rocket is a plant which grows in sand just above high tide mark on beaches in California. A biologist wondered why it was not found in the meadows next to the beaches. Which of the following hypotheses could NOT explain the absence of sea rockets from the meadows?

1)   In the next stage of succession it will invade the meadows and become dominant there.
2)   It can only grow in loose sand with very little organic soil.
3)   Beach conditions, such as spray and direct sunlight, are essential for sea rocket growth.
4)   Rabbits eat sea rocket sprouts anywhere except on the beach.

5) Sea rocket seedlings can't grow where other plants are already growing because of the shade created at ground level by the other plants.

34. The biologist dug test plots 10 feet square in the meadow, filled them with beach sand and planted sea rocket seeds in them and on the beach. Which of the hypotheses in the previous question was he testing?

   1) 1 only.
   2) 2 and 4.
   3) 3 and 4.
   4) 2, 4 and 5.
   5) 2, 3 and 4.

35. The biologist found that sea rockets grew better in the test plots of Question 34 than on the beach. Furthermore, they grew even better when seeds were planted in test plots made by spading the meadow soil so that no other plants were growing in the plot. He concluded that hypothesis 5 in Question 33 was the best. Which of the following studies would he most likely do next to strengthen his conclusion?

   1) Grow sea rocket seedlings in wire enclosures to keep out rabbits.
   2) Grow sea rocket seedlings in various light intensities.
   3) Plant seeds of meadow plants on the beach.
   4) Plant sea rockets in test plots of meadow soil on the beach.
   5) Go fishing. After all, he reached a conclusion.

36. Which of the following assumptions did the biologist most likely make in reaching his conclusion?

   1) The light intensity on the test plot is higher than on the beach.
   2) The salinity of beach sand is reduced when rain falls on it in a test plot.
   3) The temperature is lower on the beach than in the test plot.
   4) The light intensity is lower under the meadow plants than above them.
   5) Rabbits do not go onto the beach because they would be exposed to predators.

37. Paramecia A and B have very similar food requirements. An ecologist grows these two species of paramecium together for 30 days and obtains the results shown in the graph. He concludes that the extinction of B is due to its inability to compete successfully with A. What additional experiments and observations would best support his interpretation?

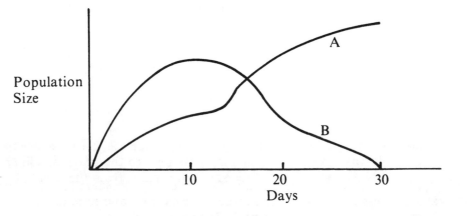

1) A is grown alone under the same conditions of food, temperature, etc. as in the above experiment. Its growth curve is the same as for A in the original experiment.
2) B is grown alone under the same conditions of food, temperature, etc. as in the above experiment. Its growth curve is the same as for A in the original experiment.
3) A and B are grown together under the same conditions and with the same results as in the original experiment.
4) When A and B are grown together for 40 days they both become extinct.
5) A is grown alone under the same conditions of food, temperature, etc. as in the original experiment. Its growth curve is the same as that of B in the original experiment.

38.   Some marine snails (species L) produce large numbers of small eggs that are shed into the sea, hatch at an early stage of development, and feed on microscopic plants. Other marine snails (species S) produce few eggs that are retained for a long time within the body of the mother snail in a brood pouch. They hatch at a very advanced stage of development, and immediately begin to feed on the surfaces of large plants like the adults. If a biologist discovers that there are approximately equal numbers of adults of species L and species S in the same area over many generations, which of the following is most likely?

1) Young cared for by parents until they reach a more advanced stage have a greater chance of survival.
2) Approximately the same percentage of young produced by each species survive to maturity.
3) Organisms which produce a greater number of offspring will likely replace organisms with fewer offspring.
4) The young of species L grow more rapidly than the young of species S.
5) Both 1 and 3 are equally likely.

39.   While studying the two species of snails in Question 38 a biologist hypothesized that the two species expend approximately the same amount of energy producing young. Which of the following is he most likely assuming?

1) Species S snails could produce as many eggs as species L snails under certain conditions.
2) An egg of species S contains more energy than an egg of species L.
3) Since species L snails produce more eggs, these eggs probably contain more energy.
4) Species S snails provide some energy to the young when they are in the brood pouch.
5) Both 2 and 4 are equally likely.

40.   A scientist proposed that humans possess a physiological mechanism which could limit population size. Which of the following would be the most effective way such a mechanism could work?

1) When women live and work in crowded conditions, estrogen production increases.
2) The more children a woman bears, the longer it takes her to become reproductively active again.
3) The greater the number of sisters a girl has, the longer the duration of time to her first ovulation.
4) 2 and 3 are equally likely.
5) All of the above are equally effective.

41.   If one or more of the physiological responses to social conditions described in Question 40 exist in humans, why are there so many people, especially in very crowded parts of the world, such as India? Which of the following is the most probable answer?

1) Humans are fundamentally different, biologically, from all other organisms, so no hormonal mechanism of population control could possibly work.

2) If any mechanism did occur in some women who are sensitive to variations in social conditions, they would have left fewer offspring, therefore this could not have persisted over many generations.
3) Because "clutch size" in humans is normally one, and seldom more than two, the mechanisms that limit family size in birds don't work in humans.
4) Humans are more selfish, less altruistic in limiting family size for the benefit of our species than are members of other species, such as mice.
5) Human reproductive behavior is not controlled by either hormonal or social factors.

42. A bird watcher observed that cowbirds would lay several eggs in a nest of *Oropendula* (a sub-tropical bird) in area A and lay only one egg in an *Oropendula* nest in area B. The bird watcher proposed that the cowbird would lay only as many eggs in the *Oropendula* nest as would be accepted by the owner of the nest. What assumptions did the bird watcher make in proposing his hypothesis?

1) Some *Oropendula* can better discriminate between cowbird eggs and their own than other *Oropendula* can.
2) The environments must differ between the two nest areas.
3) Nesting season must differ in these two areas.
4) Both 1 and 2.
5) All of the above.

43. Which of the following experiments would you do to test the hypothesis in Question 42?

1) Add several cowbird eggs to the *Oropendula* nest which contained one cowbird egg (area B) and observe what the *Oropendula* does with them.
2) Remove several cowbird eggs from the *Oropendula* nest which contained several cowbird eggs (Area A) and observe what the *Oropendula* does with the remaining eggs.
3) Remove the cowbird egg from the *Oropendula* nest which contained one cowbird egg (area A) and observe what the *Oropendula* does with the remaining egg.
4) Either 1 or 2 would test the bird watcher's hypothesis.
5) Either 1 or 3 would test the bird watcher's hypothesis.

44. The bird watcher of Questions 42 and 43 observed that young cowbirds which developed in the nest in area A removed parasites from the young *Oropendula*, while the single cowbird which hatched in area B did not remove parasites from the *Oropendula* young. The bird watcher hypothesized that the reproductive success (number of young which reach breeding age) of the *Oropendula* in each area would be similar. Which of the following is the bird watcher not assuming in stating this hypothesis?

1) The young *Oropendula* in area B are subject to parasites like the *Oropendula* in area A.
2) Parasitism can lead to reduced reproductive success if unchecked.
3) The young *Oropendula* in area B have no means of countering parasitism.
4) The young *Oropendula* in area B remove parasites from each other.
5) Cowbird hatchlings compete for food with *Oropendula* hatchlings in both areas.

45. Consider two hypotheses:

A. Each bird in a population lays as many eggs in a clutch as it can lay that season.
B. Each bird in a population lays only as many eggs in a clutch as it will be able to feed when they hatch.

Certain birds lay only one egg in a clutch. If this egg is removed, the bird lays one more. This observation:

1) Supports hypothesis A.

2) Supports hypothesis B.
3) Is consistent with both hypotheses.
4) Contradicts hypothesis A.
5) Contradicts hypothesis B.

46. The same birds mentioned in Question 45 were the subject of an experiment. An extra egg was placed into each nest, raising clutch size to two. The number of young which were raised to leave the nest was 75% greater than the number raised in control nests which had only one egg. These results:

1) Support hypothesis A.
2) Support hypothesis B.
3) Are consistent with both hypotheses.
4) Contradict hypothesis A.
5) Contradict hypothesis B.